The Ethics of Intensification

The International Library of Environmental, Agricultural and Food Ethics

VOLUME 16

For other titles published in this series, go to
www.springer.com/series/6215

Paul B. Thompson
Editor

The Ethics of Intensification

Agricultural Development and Cultural Change

Springer

Editor
Paul B. Thompson
Michigan State University
East Lansing, MI
USA

ISBN: 978-1-4020-8721-9 e-ISBN: 978-1-4020-8722-6

Library of Congress Control Number: 2008931061

Printed on acid-free paper

9 8 7 6 5 4 3 2 1

springer.com

Dedicated to Don Ihde, my Doktorvater, who escaped Kansas.

It has been my lot to ensure that he does not entirely escape the farm.

Acknowledgments

This volume grew out of a workshop sponsored and co-funded jointly by the Food and Agricultural Organization of the United Nations, Imperial College London and Michigan State University. Peter Kenmore and Clive Stannard were the key individuals at FAO. Jeff Waage provided administrative and funding support from Imperial College London, while Andrew Dorward helped enormously with organization and local arrangements at the Wye Campus. Dan Clay, Director of International Agricultural Programs at Michigan State, provided funds for travel by MSU faculty. Other individuals played key roles as the project moved from a workshop to a book. John Otieno Ouko dedicated a semester of work to interacting with authors and helping prepare the prospectus. At Springer, Fritz Schmuhl and Marion Wagenaar provided help and encouragement. Final preparation work at MSU was done by Ellen Link. Julie Eckinger was indispensable to maintaining smooth communications and transitions as the book moved through each of these stages. Without her, this would have been impossible.

Financial support for all of these people was provided through funds made available to the W.K. Kellogg Chair in Agricultural, Food and Community Ethics at Michigan State University through a gift from the W.K. Kellogg Foundation of Battle Creek, Michigan. Errors and omissions remain the responsibility of the editor.

Paul B. Thompson

Contents

Contributors

Richard Bawden is Dean Emeritus of the Hawkesbury Agricultural College, now a component of the University of Western Sydney in New South Wales, Australia. He is recognized as one of the foremost authorities on systems thinking in agricultural development.

Lawrence Busch is University Distinguished Professor in the Department of Sociology at Michigan State University and Professor at the Center for Economic and Social Aspects of Genomics at Lancaster University. He is the author of a number of publications on the sociology of the agricultural sciences.

Luis Camacho is Emeritus Professor of Philosophy at the University of Costa Rica. He has written widely in the philosophy of science and development ethics.

Andrew Dorward is Professor of Development Economics in the Department of Financial and Management Studies, School of Oriental and African Studies, University of London. He has published widely on the economics of agricultural development.

Nigel Dower is Honorary Senior Lecturer in Philosophy at the University of Aberdeen. He has written extensively in development ethics and global ethics and was President of the International Development Ethics Association 2002–2006.

Stephen L. Esquith is Dean of the Residential College in Arts and Humanities at Michigan State University. He has written widely in political philosophy and designed a study abroad program focusing on ethics and development.

David Fraser is Professor in the Animal Welfare Program at the University of British Columbia, and holds a cross-appointment in the W. Maurice Young Centre for Applied Ethics.

Rachael Jamison coordinates the Green Building Program for the Washington State Department of Ecology. She was formerly an associate with the Organic Food Program in the Washington State Department of Agriculture.

Michiel Korthals is Professor and Chair of the Applied Philosophy group at Wageningen University and Research Centre. He is the author of *Before Dinner* as well as many articles on ethics, sustainability, agriculture and nutrition.

John Otieno Ouko is a graduate of the University of Nairobi who has written a number of papers and reports on human rights issues in Africa. He is currently pursuing Ph.D. studies in philosophy at Michigan State University, focusing on ethics and development.

Clare Palmer is Associate Professor in Philosophy and Environmental Studies at Washington University in St. Louis. She is President of the International Society for Environmental Ethics and has published widely in environmental ethics.

John H. Perkins was, before his retirement, a member of the environmental studies faculty at The Evergreen State College. His original training was in biology and history of science, technology, and environment. He is the author of several books on the history of agricultural research and is currently working on a new book on energy.

Peter Sandøe is the Director of the Danish Centre for Bioethics and Risk Assessment and Professor of Bioethics at the University of Copenhagen. He is, together with Stine B. Christiansen, the author of *Ethics of Animal Use* and has published many articles on ethical issues relating to animals, biotechnology and food production.

A. Allan Schmid is University Distinguished Professor Emeritus in the Department of Agricultural, Food, and Resource Economics at Michigan State University. He is the author of many books and articles on the institutional approach in economics.

Clive Stannard was, before his retirement, Assistant Secretary of the Commission on Plant Genetic Resources for Food and Agriculture at the Food and Agriculture Organization (FAO). He is by training a social anthropologist, with an African focus, and a historian.

Paul B. Thompson holds the W. K. Kellogg Chair in Agricultural, Food and Community Ethics at Michigan State University. He has published many books and articles on ethical issues in agriculture and food production.

Introduction

The Ethics of Agricultural Intensification: An Interdisciplinary and International Conversation

Paul B. Thompson and John Otieno Ouko*

Global agriculture faces a number of challenges as the world approaches the second decade of the third millennium. Predictions unilaterally indicate dramatic increases in world population between 2010 and 2030, and a trend in developing countries toward greater consumption of animal products could multiply the need for production of basic grains even further. Although global food production in 2000 was estimated to be adequate for the existing population, hunger and malnutrition are persistent problems that have led decision makers to recognize that increasing food production in specific regions may be the most effective way to address food security for impoverished peoples. At the same time, there will need to be policy adjustments that improve poor people's access to current food supplies without simultaneously undercutting the ability of local producers to obtain needed cash income. What is more, the uncertain effects of global climate change on agricultural ecosystems complicate planning for this process, while poorly understood processes of globalization create additional unknowns from the side of social systems. In short, despite surpluses in many parts of the developed world, finding ways to increase food production on both selected regional and a total global basis remains a priority for many farmers, policy makers and agricultural researchers.

There are essentially two routes to adjusting total food production on a regional or global basis. Extensive approaches focus on the total amount of land being used to support agricultural production, while intensive approaches focus on technology and the organization of labor being deployed on a fixed amount of land. A commitment to maintaining existing farming methods and ways of life implies that adjustments will be made by bringing additional land into production. But with total arable land coming into increasingly short supply across the globe, it becomes critical to examine the ethical issues that are raised on a global basis by following the route taken in most industrialized economies: using technology to irrigate, fertilize, and otherwise manage agricultural production, allowing broad and persistent increases in per hector yields. The essays in this volume start that conversation on ethical issues.

---

* Paul Thompson, is the W. K. Kellogg Chair in Agricultural, Food and Community Ethics, Michigan State University. John Ouko is a doctoral student, Department of Philosophy, Michigan State University

P.B. Thompson (ed.) *The Ethics of Intensification*,

Understanding Intensification

As Clive Stannard's contribution to this book discusses in some detail, the Food and Agriculture Organization (FAO) of the United Nations began to recognize the need for an explicit and planned effort to address ethical issues in agriculture more than a decade ago. The theme of agricultural intensification was one of several that FAO selected for coverage in their series of ethics papers. This volume grows directly out of efforts undertaken at FAO. Chapter 1 reprints the FAO Ethics Paper on agricultural intensification, while several other chapters are structured as the author's reaction to it. But the very idea of intensification is problematic. As will be discussed at several intervals throughout this volume, framing the future of world agriculture in terms of a choice between intensification and continued expansion of the land base dedicated to food and fiber production is itself an ethically contentious way to proceed. On the one hand, the future of agricultural production will almost certainly involve both of these strategies. New lands can be brought into agricultural production in some parts of the world, and climate change may introduce additional possibilities for adjusting land use. On the other hand, some observers would argue that describing changes in terms of intensification only obfuscates the ecological, economic and technological drivers of change. When obfuscation conceals exploitation, it becomes ethically problematic in itself.

In fact, the decision to frame this issue in terms of intensification was made within FAO itself. It was made against the backdrop of widespread recognition of the need to confront a division of opinion within FAO about the legitimacy of directed change in farming technology and in the policy environment guiding farmer decision making. The question might be posed thusly: Is it ethically legitimate for an agency such as FAO to undertake projects intended to change the way that people farm? If so, what ethical principles should guide these projects? If not, what justifies a posture of failing to do things that could be understood as an ethical imperative to aid poor and hungry people? While there are indeed issues that are unique to the kind of international organization that FAO is, the real interest here was in the ethics of *any* agency undertaking such activities, whether it be a government, an inter-governmental organization, a charitable foundation, a civil society group, or indeed, a for profit corporation (or a well-endowed individual). In other words, the goal was to spark a conversation about the ethics of the act, not the peculiar characteristics of the agency that undertakes it.

Parties within FAO and its constituencies recognized that the very word 'intensification' was a potential source of confusion and debate. As noted above, ever since antiquity farming and pastoral peoples who needed more food and fiber have gotten it by expanding their land base. That is, by simply increasing the amount of biological input they achieved more output in the form of consumable goods. But the agricultural producers of antiquity also deployed other means, such as using more water (irrigation) or applying nutrients, usually in the form of animal manures. What is more, ancient farmers recognized that adding more labor in the form of careful hand weeding, better tillage or improvements to drainage can also add a boost to the amount of food and fiber commodities they harvest each year. All of these strategies can be construed as ways of increasing inputs in order to increase outputs. All of them are components of traditional farming systems, yet only the first need necessarily involve an increase in the land base on which a given community relies. So defining intensification as an increase in commodity yields per acre can be a very misleading indicator, and one that fails to capture what this conversation needs to be about.

But other terms also have the potential to mislead and to take the conversation off course. For some, the debate centers on "the Green Revolution," understood as the application of modern agricultural science to the farming methods being deployed by farmers in the developing South. But for others, this phrase refers to a specific initiative launched by the Rockefeller Foundation after World War II, gradually developing into the Consultative Group on International Agricultural Research (CGIAR). While the essays that follow do not neglect this initiative or the debates over its legacy, a narrow focus on this activity would not only be somewhat backward looking, it would also neglect the profound impact that national governments and private firms have upon the mix of agricultural production methods, tools and techniques. The term 'industrialization' suffers from similar but perhaps even more extreme difficulties. Does it imply that decision making will be in hands of multi-national corporations? Certainly many of those who advocate some directed program of scientific research, technological innovation and adaptive management (i.e. intensification) want no part of decision making by multi-national corporations. (Though, of course, others would concede that they will play a role, in any case.) These issues on the role of private capital broach yet another debate, discussed below as well as throughout the volume.

Though there remain important topics to debate within these contested issues, there was a juncture within FAO's internal discussions on the ethics project where it was decided that it was more important to go ahead and begin this conversation than to continue a seemingly endless search for the perfect terminology with which to have it. Such is and always has been the inherent nature of a philosophical debate. Terminology is imperfect and inherently misleading, but the importance of the issues at hand demands an exchange of views, nonetheless. The essays and title of this volume operate within the legacy of that framing decision, and while all the authors in the volume recognize the conceptual limitations and potentially controversial implications of a phrase like "the ethics of agricultural intensification," all have accepted the need to go ahead and have this conversation, even while we continue to debate the words and concepts with which we will have it.

As the chapter by John Perkins and documents, debates over intensification strategies in agriculture have occurred for many years. Since World War II they have often been framed in terms of the success and failures of the Green Revolution, however broadly or narrowly we understand that term, or in terms of social and environmental costs associated with intensification strategies in the developed world. Until comparatively recently, these debates were not framed in terms of ethics. That is, although protagonists clearly held values, their values and value orientations were often implicit and were almost never articulated in the terms of an explicit ethical perspective. Thus, while the word 'intensification,' gives rise to one set of difficulties, the word 'ethics' gives rise to another.

Understanding Ethics

When discussion of benefits, risks or human rights was introduced into the debates reviewed by Perkins and Jamison, these concepts were almost never defined or specified as involving systematic ways to approach and understand social goals or normative expectations. The 20th century debates over intensification, in short, exhibited little cognizance of the way that philosophers and social theorists working in ethics and political philosophy had attempted to frame concepts such as benefit, risk or human rights within comprehensive theories. Neither did they engage with emerging philosophical discussions on the ethical standing of the natural environment or non-human animals.

But to insist that such debates *should* utilize constructions and ideas developed by philosophers and social theorists is also to make a very contentious statement. One source of controversy resides in the training and mindset of people who have had formal training in agriculture and applied social science disciplines. Studies by Glenn L. Johnson (1976, 1982), Jeffrey Burkhardt (1997, 1999) and Robert Zimdahl (2006) have concluded that the education and practice of professionals such as those that staff organizations such as FAO, the CGIAR, agricultural universities, national research institutes and ministries of food and agriculture is deeply imbued with the philosophy of *positivism*. Philosophical positivism is a belief that scientific method and technical innovation can be broadly applied in solving social problems. Positivists since Auguste Comte (1798–1857) have denigrated the ability of humanistic, religious and interpretive traditions often associated with the word 'ethics', but the specific version of positivism referenced by Johnson, Burkhardt and Zimdahl evolved from the writings of physicist Ernst Mach (1838–1916) and was given definitive formulations in Vienna and in the United Kingdom during the decades before World War II. This philosophy held that the subjective nature of moral beliefs made them inherently incapable of demonstration and unsuited to rational disputation. Ethical claims were said to be "meaningless," hence ethical values have no place in science The view was short lived in academic philosophy, as it was quickly recognized that the statement "Ethics have no place in science," is itself an ethical norm for the practice of science. To propose a philosophy of science based on such a claim would be internally incoherent. Nonetheless, the view did clearly continue to have influence amongst the scientific disciplines long after the war. Johnson, Burhardt and Zimdahl each argue that this influence lingers in the agriculture disciplines today.

However, the positivist's tendency to quash ethical debate receives support from purely practical considerations, as well. For one, claiming that a practice is "unethical," can be counterproductive when the goal is to get it changed. Such talk can make people defensive and can wrongfully impugn the motives of the well-intentioned.

Another problem lies in the need to engage in lengthy terminological clarification (a problem made evident by the tone of this introduction). While almost everyone thinks that they are themselves ethical people, discussing ethical norms and values is surprisingly complicated. Terminological misunderstanding begets contention, and conversations become both wearisome and acrimonious. For these inherently sound reasons, many people with practical problems that need solutions avoid discussions about ethics whenever possible.

Academic and scholarly studies in ethics are, in fact, partially a response to these very problems. Philosophers and social scientists specializing in ethics develop vocabulary for more precise and efficient ways to discuss right and wrong. Theories of ethics depersonalize debate so that a given view of what is right or wrong in a given situation comes to be understood as owing to standard or stylized approaches to the question of determining right and wrong, rather than to the ethical (or unethical) character of the agents involved in the situation. One contribution that academic ethics can potentially make to debates over intensification lies in the creation of a standardized and depersonalized vocabulary in which to discuss the competing goals, constraints and approaches that shape both thinking and action in pursuit of agricultural improvement. But while this hope of developing such a vocabulary serves as the primary rationale for this volume, it also points toward more intractable problems with our approach.

While every culture and linguistic tradition has some mode for articulating the idea of ethics and for discussing right and wrong, the modality of academic and scholarly ethics has been dramatically shaped by the history and institutional environment of the European university system. Of course, the same can be said of the agricultural sciences. Yet while the methods and findings of the sciences have come to have almost worldwide acceptance, local, national and regional traditions of ethics continue to provide authoritative sources of insight into matters of right and wrong that cannot be "trumped" or overridden by the academic theories that are the product of Western philosophy departments and the social sciences. Thus while farmers continue to know things about their local soils that have never been fully codified by agricultural researchers, this fact is not thought to compromise the authority with which a soil scientist articulates the basic principles of the discipline. Ethics does not enjoy this kind of status, and very few ethicists working in Western university environments would seek to achieve it.

FAO papers and research efforts on ethics have made attempts to reflect the broad cultural and religious diversity found among member nations and farming peoples around the world (Bhardwajet al. 2003; Macer et al. 2003). Unlike these efforts, the paper on intensification was specifically intended to draw upon ethical theories and approaches as they are currently understood by academic philosophers and social scientists. This volume, too, has been conceptualized in that mold, though every contributor recognizes that this represents a limitation on the very idea of ethics. Indeed, the last quarter of the 20th century saw widespread change within academic ethics, as numerous scholars challenge the Euro-centric perspective of the leading traditions in ethical thought. Theorists of race and colonialism offered penetrating critiques of the way that academic ethics valorized power and property, and feminists suggested an implicit "phallogocentrism" or privileging of the male perspective. Thus it is quite reasonable to argue that the philosophical tradition informing both the FAO Ethics Paper on agricultural intensification as well as the commentary articles that follow is itself undergoing radical transformation and revision.

It would have been inherently desirable to involve a much broader range of cultural as well as theoretical perspectives in the exercise that has culminated in this volume. Well developed philosophical traditions exist in Asia that are completely missing from this volume. Latin American philosophy is, like that of the North, deeply influenced by the European tradition, yet it has developed its own style and orientation, represented here by only a single contribution from Luis Camacho. African philosophy has made strides toward the recovery of an authentic scholarly voice silenced during the colonial era, but again, the present volume includes only one example: John Ouko's adaptation of critiques of the Green Revolution. One could also argue that many voices from the European tradition itself are also missing. In fact, the limiting factors that eventuated in the less desirable outcome that this volume represents are practical rather than ideological. The contributions needed to be in English, for example, and there were no funds for translation.

For the most part, however, the shortcomings must be laid at the feet of the editor. Although I had substantial help and advice throughout the planning and preparation of this volume, there eventually came to be time when the obvious need for more far flung and diverse perspectives was overcome by my own limited imagination, contacts and energy, especially as the imperative for bringing the work that had already been contributed by others to fruition. Hopefully others will regard this volume as the beginning of a conversation, and will take up pen and paper to join in.

Ethics and Agricultural Intensification

As such, the perspectives represented within these covers strike a balance between diversity and systematic coverage. Chapters by Luis Camacho, Stephen Esquith and Nigel Dower each situate this volume within the emerging tradition of "ethics and development," while Claire Palmer, Michiel Korthals and Peter Sandøe each employ recent ideas developed in environmental philosophy. The phenomenon of intensification is reasonably seen as a feature of planned development, whether it is viewed as a logical extension of publicly funded scientifically based research efforts that had their origins in England and Germany or of private sector innovation and economic growth. As noted already, and documented in this volume by contributions from Richard Bawden, Allan Schmid and Lawrence Busch, intensification has been widely debated by agricultural specialists for some time. Is there anything that philosophers can actually add to that debate?

While readers are invited to answer that question for themselves by reading the chapters, it may be useful to frame that reading with just a few more anticipatory remarks on the overall problem. In 1899, Karl Kautsky published *Die Agrarfrage*, translated into English only relatively recently as *The Agrarian Question* (1988). Kautsky argued that innovations in agricultural production and post-harvest distribution were having the effect of allowing those who controlled the technology behind the innovation to gain control of the benefit stream associated with agriculture. In short, farmers do not benefit from innovation, even though they are compelled by competitive pressure to adopt innovations. Instead, innovations are actually a means for non-farmers to accumulate or co-opt some of the returns that would have gone to farmers and to keep it for themselves. Kautsky's argument expanded upon Karl Marx's analysis of the way that applying industrial principles to agriculture resulted in inevitable exploitation and depletion of the inherent productivity of land (Marx 1906).

The basic elements of this analysis have been repeated many times over the last 100 years, sometimes with direct reference to Kautsky's work (see Goodman and Watts 1997), and sometimes not (see Cochrane 1983). At the same time, the mere association of this argument with names like Marx and Kautsky may have been all the reason that ideologically inclined agricultural researchers and government officials needed to repress it. The upshot is that the debate over agricultural intensification has ideological overtones that one neglects at one's peril. Those who have accepted the basic tenets of Kautsky's analysis occupy a range of stances along the political spectrum, but they all tend to believe that unless specific measures are taken to insure against it, technologically induced efforts at intensification will benefit the rich and powerful at the expense of primary agricultural producers (who, outside the developed West, tend to be weak and poor). They also tend to see those who think that science and technological innovation are inherently progressive as at best naïve and at worst as scheming to gain at the expense of others.

Those on the other side also occupy a range of positions along the political spectrum. Some positivist-inclined scientists do indeed see innovation as inherently progressive. Some see the socialist argument as a ruse to concentrate government power at the expense of individuals. Others see increases in total food production as inherently favoring the interests of the poor, and hence see the benefits of intensification as off-setting losses to those farmers who, they admit, suffer as a result. Still others are willing to discuss ways that more sophisticated approaches might either compensate losers, or at least minimize their losses. Thus it is not as if there is no middle ground in the intensification debate. Marx and Kautsky themselves were hardly opposed to technical innovation, though they were not sanguine about innovation under the basic principles of a capitalist economy.

There are numerous echoes of this fundamental debate sounding throughout various chapters in this book, but the basic point of an ethics-oriented approach has been to raise questions that are somewhat orthogonal to it. Specifically, an ethicist asks whether one sees the potential for diverting benefits from primary producers as politically justified or not, *on what normative basis does one judge?* If one takes the view that what ultimately matters is the way that change affects the welfare or quality of life for farmers (and possibly non-farmers, as well), the ethical debate will proceed in one direction. If one thinks that what ultimately matters is the fairness (or unfairness) with which these shifts and changes play out, a very different set of considerations will emerge. Or perhaps one thinks that the cultures and ways of life lived by these farmers (and possibly others, as well) are constitutive of what ultimately matters, so that changes must be evaluated in terms of the violence that they do to the underpinnings of a worthwhile life.

The debate over intensification tends to entangle the values and arguments that have been mustered in support of one perspective on the agrarian question (or leveled against another perspective) with a perceived rooting-interest. Thus if one is "for" farmers, one sees any argument leveled in support of non-farmer interests as inherently tied to those interests, and if one is "against" government, one sees any argument suggesting a role for government as antithetical to one's rooting interest. In contrast, an ethics-based analysis will tend to lump the pro-farmer advocate who cites a Kautsky-type argument to show that farmers are losers and the pro-capitalist advocate who says that low food prices for the poor offset these losses into the same camp. It's their reasoning rather than their partisanship that puts them there. Both are using ethical criteria that stress the outcome or endpoint of an innovation process and that emphasize welfare or income. This would stand in contrast to someone who sees even a positive outcome for welfare and income as ethically problematic when it is brought about in a coercive manner.

The point can be made in a more direct (if also more contentious) manner as follows: For some observers of agricultural intensification, the long squabble over capitalism is everything. Those who fail to declare an allegiance in that battle at the outset are presumed to be sleeping with the enemy. Given the sensitive nature of the word 'ethics,' such partisans are particularly inclined to be suspicious of an approach that neither decries nor praises capitalist institutions in an overt manner. Thus, although ethical theories do indeed intersect with this debate in powerful ways, the primary thrust of philosophical ethics does *not* set out to define good and bad in terms of an allegiance or opposition to capital.

One last anecdote might also help. I serve on the Ph.D. committee for a sociology student who uses the expression "social justice" to indicate a particular ethical orientation, advocated by some and opposed by others. I tell him, no, social justice is an ethical problem that different ethical theories will address and analyze in different ways. But my admonishment is to no avail. Arguably, my student's indoctrination in social movement theory leads him to see normative commitments solely in terms of the way that they bind people together into groups that are more or less effective in pursuing a set of common goals or interests. He is so intently focused on the behavioral and political implications of expressing allegiance to a phrase (social justice) that he is actually rather uninterested in whatever it is that people might take the phrase to mean, or the reasons they adduce for seeing social justice in one way rather than another.

This is not to suggest that the excellent social science dedicated to the agrarian question is irrelevant to ethics. Understanding how efforts at intensification both create and then effect economic and political interests is a crucial component to any fully developed understanding of agricultural ethics. Understanding the way that technical innovations contribute to processes of domination and resistance is critical to ethics. But an ideological position based solely on an analysis of who wins and who loses is rather akin to a rooting interest in an athletic contest. Ethics is not about being a fan (or an antagonist) of Manchester United or the New York Yankees, nor is it (solely) a matter of having a rooting interest in the fate of capital *or* the poor.

Nearly a Decade in the Making

So what would a deliberate attempt to stimulate conversation between agriculturalists practically engaged in strategies of intensification and philosophers attentive to the challenges of ethical thinking look like? In partial answer to that question, Michigan State University and the Wye Campus of Imperial College, London convened a small group of philosophers and practitioners (themselves representing an assortment of disciplines in the agricultural sciences) in January of 2004 for two and half days of conversation. The session was structured as an open-ended give and take in reaction to an ethics paper then nearing completion for FAO. Several participants prepared direct commentaries on the paper, while others reacted to these commentaries. The results of the meeting prompted the design of this book. The original FAO paper is reprinted as Chapter 1, and several participants have contributed papers that develop ideas expressed at the meeting in their chapters. Other chapters are from authors who take the theme of ethics and intensification beyond the discussion at the meeting, extending the conversation into new realms.

The Wye conference had its origins in the FAO ethics process described in some detail by Clive Stannard (Chapter 2). I had been contacted four years earlier by Peter Kenmore of FAO, and our conversations resulted in my being retained under a consulting agreement with FAO to provide advice in drafting some internal documents as well as the FAO Ethics Paper on agricultural intensification. I was joining a process that had already been underway for at least a year. A fair amount of text for the paper had already been written when I came on board, but one specific request was to develop an approach to the subject that would reflect how each of the main schools of thought in philosophical ethics might approach the question of intensification differently. For this task, I adapted material from a PowerPoint presentation that I had been giving to agricultural groups since the early 1990s, and worked closely with a number of groups at FAO to be sure that the resulting paper reflected the intensification debate as they understood it. I turned over the text of a penultimate draft in 2002, and it was subsequently edited both for style and content, then translated into four languages before being published as FAO Ethics Paper No. 3 in 2005.

I showed my PowerPoint and presented the paper at the Wye conference. That resulted in a number of the contributors to this volume referring to it as "Thompson's paper," and I cannot guarantee that all of these references have been corrected in the present volume. While I am perfectly happy to be associated with the result of this long FAO process, calling it "Thompson's paper" is simply not correct. Perhaps 70% of the current text originated on my word processor, but at best half of that represented my own work. Even that portion I contributed simply represents the collective judgment of the academic philosophy community as opposed to something for which I can be understood to deserve credit for having created. The document which follows this introduction as Chapter 1 is an FAO paper, issued without specific authorship, widely disseminated as such, and available on the FAO website, It is not my work.

The Wye conference was co-funded by MSU and Imperial College, which received some contract support from FAO. MSU's contribution consisted in supporting travel for four individuals to participate (hence the predominance of MSU authors in the current volume). The budget was small and the planning horizon was short. There was little advance publicity and the invitation list was constructed by trying to reflect some degree of both disciplinary and international diversity, though in the end it was availability that may have dictated who participated. The conference did include a more culturally diverse participant list than does this volume. It concluded with a lively session focused on next steps. Some of the points agreed to in that concluding conversation formed the basis for Andrew Dorward's concluding chapter of this book. One further point of agreement was that the work already invested by a number of people who had made presentations should not be lost, and that a book should be produced. This, of course, is that book.

The original idea was 4 long chapters each contemplating a dimension of the original FAO Ethics Paper, with 2 shorter chapters of commentary on each of those 4, leading to 12 chapters overall. That, plus the original paper, Rachacl Jamison, Stannard's overview and Dorward's afterword would have been the volume. However, a number of the people who made contributions at the Wye conference elected not to contribute to the book. I was thus left without a full complement of chapters and found myself in pursuit of appropriate additions, a task that took almost two years. In the meantime, chapters that had been written early in the process were on the verge of becoming dated. Of the current chapter authors, Rachael Jamison, John Ouko, David Fraser and Richard Bawden did not attend the Wye conference. While a collection of authors more representative of the global diversity of perspectives might have been desirable, I am especially grateful for these contributions that allowed me complete the book in a not entirely untimely fashion. The primary casualty of this delay is Stannard's chapter, which references a number of events in future tense that have (or should have) by now transpired. In fact, not all of them have transpired; the Ethics of Food Safety consultation he mentions was redefined as a "Readings in Ethics" paper for reasons that are not altogether clear to me. But given that Stannard is now enjoying a well deserved retirement and no longer able to speak for what happens at FAO, I have decided to print the chapter pretty much as we received in the autumn of 2004. Readers are advised not to take all of his predictions at face value.

One thing that has been retained from the Wye conference is the idea of a conversation on ethical issues and approaches. The aims of the book are to be exploratory and experimental, rather than decisive or academic. Thus some chapters explore ways to articulate and critique strategies and rationales of intensification in explicitly philosophical terms, while others articulate challenges of intensification (and thinking reflectively about it) in more practical and institutional terms. Still other chapters reiterate the model from the MSU/Wye conference, reacting to the statements and views of others and probing new lines along which the conversation might be profitably pursued. The result is predictable in some respects: Philosophical articulations are often insufficiently practical for practitioners, while more practical strategies are insufficiently philosophical for the philosophers. Yet like the participants at the MSU/Wye meeting, all of the contributors to this volume believe that narrowing the gap between philosophical ethics and agricultural policy and practice is worth doing. All believe that there are specific steps that can be taken to do this, and none express skepticism about the importance of taking those steps.

The Plan of the Book

The substantive chapters begin with the paper "The Ethics of Sustainable Agricultural Intensification," prepared for the Food and Agriculture Organization (FAO) series on ethics. It focuses on the way ethical concepts and ethical terms can be used to articulate normative dimension whenever episodes of intensification occur. The paper provides a basic conceptual framework in ethics that will aid in the articulation and justification of norms for intensification, but does not defend a particular set of answers to the ethical questions that arise in intensification. It summarizes how the three main ethical approaches, consequentialist, rights-based, and virtue based theories, can be used to answer three important questions as regards agricultural intensification: When is intensification ethically good? How should burdens and benefits of (intensification) be distributed? And who is responsible to see that intensification occurs? The paper concludes that it is critical to articulate the ethical rationale for undertaking projects in intensification in terms that draw upon each of the approaches. Omitting any one of these ways of framing ethical issues results in a weakened capacity to articulate, debate, and ultimately perform ethical responsibilities that may arise in connection with the imperatives for intensification. The FAO's illustrated version can be accessed on line at http://www.fao.org/docrep/007/j0902e/j0902e02.htm

In Chapter 2 Clive Stannard gives an extensive description of the steps Food and Agriculture Organization of the United Nations (FAO) has taken to approach the application of ethical perspectives to food and agriculture. He also highlights some of the questions FAO has been asking in developing its ethics program. The main steps FAO has taken are: the formation of a panel of eminent experts on ethics in food and agriculture; and the production of publications on ethics in food and agriculture (***The FAO Ethics Series***). The panel of eminent experts promotes reflections upon ethical issues arising out of food production and consumption. Issues relating to the interests of future generations are also taken up by the eminent experts. FAO Ethics Papers address the ethical components of farming, how best to relate ethical discourse to other modes of discourse such as economic and development theory, how intensification relates to social justice, and whether harm must be done to some group, or to the environment, because the long-term benefits justify this. The series "Readings in Ethics" complement the ethics papers with additional reflections that represent the perspective of specific individuals or groups.

John Perkins and Rachael Jamison summarize the history of some particularly interesting cases of agricultural change in Chapter 3. Their main argument is that understanding the dynamics of science and technology is necessary but not sufficient to understand agricultural change. Other political, economic, cultural, and ethical factors are also important. Scientists need to understand the social and political environment of agricultural change when they plan their work, since changes in agricultural practices affect not only the scientist and the farmer but also the entire human population. The social context is particularly important when it comes to intensification. Inventors and innovators of agricultural change have seldom given much attention to the complex consequences of change and the ethical implications of technology used in agricultural intensification. Perkins and Jamison point out the ethical implications as regards mobilization of water sources, use of synthetic nitrogen fertilizers, energy supplies, some uses of synthetic pesticides, and transgenic plants. Whereas intensification has been key to producing enough calories to potentially feed the world population, agricultural scientists are faced with a challenge: how to maintain stable yields without jeopardizing agriculture or earth's resources. The issues of prior informed consent, democracy, public participation, individual and national political power, human health, and the environment complicate the choices that scientists and their supporters must make. The future prosperity of people and all other species depend on the choices made.

Chapter 4 is entitled "One Hundred Years of agricultural Intensification: A Personal History of Unanswered Ethical Issues – 1890–2004." Allan Schmid examines the ethical implications of a series of farm management choices made by his family in the process of intensification over the 100-plus year period. He notes that intensification led to specialization and need for more capital which many farmers did not have thereby being forced to leave farming. Also, with the coming of new seeds ethical issues were/are involved in whether property rights should apply to seeds at all as well as whether farmers should have the right to replant purchased seeds. Intensification has also led to higher yields but it has also led to the disappearance of some species of plants and animals. Even antibiotics fed to animals for disease control can be dangerous to consumers of animal flesh. Similar arguments apply to other areas related to intensification like energy use, labor, soil conservation, land inheritance processes, and cultural change. Schmid argues that efficiency (which intensification leads to) cannot be a guide to ethics. Efficiency depends on whose interests are a cost of production. Therefore the fact that many of us respond to available opportunities does not mean that our choices are not affecting others. We have an ethical responsibility to forgo some opportunities or to participate in the debate over the rights structure that determines our opportunities.

Many see the history of agriculture as a constant battle against famine, hunger and undernourishment. Michiel Korthals notes in Chapter 5 that Perkins and Jamison also stress the idea of increasing yield in order to feed the growing world population, but they add that consumers have a lot to say on the type of agriculture and food production, and ought to have some influence on the course of agriculture taken. A case in point is the use of biotechnology in rice production, which aroused discontent in many countries. This example and many others bring Perkins and Jamison to the conclusion that the wider public cannot be left out with respect to important agricultural innovations. On this point Korthals agrees with Perkins and Jamison, but he also identifies a second battle in the history of agricultural intensification, which Perkins and Jamison either did not see or did not talk about: the battle between types of agriculture and the way a dominant type of agriculture is used as a tool in structuring society, both culturally and politically. Agriculture is not only about higher yields, but also about the type of foodstuff to be produced, distributed, sold, bought, prepared and consumed. It is also about the food styles preferred by different people. Thus Korthals sees the history of agriculture as a continuous battle fought between parties that were already represented in agriculture and others that weren't. The history of food production is a thus an ethical battle about which food choices should be represented.

Luis Camacho examines agricultural intensification from the perspective of development ethics in Chapter 6. Camacho has taken a closer look at the historical relationship between technology and our ethical ways of life as they have developed unevenly and ambiguously over time. Agricultural intensification increases productivity in development processes yet technology is not always a good thing. Therefore, assessment of the consequences of such intensification for people and the environment are very important in development ethics. Potential victims of new agricultural technologies that intensify production should have a voice in the design as well the use of these technologies to avoid bad and evil consequences. Moreover these harmful consequences can extend to social relations and cultural identity, not just economic well-being. But without the secure cultural identity that is put at risk by some new technologies, some groups may not be able to participate effectively in the technological design process. For this dialogue between traditional culture and industrial culture to result in both more efficient and appropriate technologies for agricultural intensification, there must be a method for balancing the contributions of both locals and outsiders. Development ethics facilitates a dialogue among different groups willing to share factual experiences and moral insights in the struggle for human survival. Because of its unique role in such a survival, the answers given to questions arising from agricultural practices in general and agricultural intensification in particular are of vital importance.

Stephen Esquith appraises Camacho's "ethic for development" – which they also refer to as "development ethics" in Chapter 7. Esquith agrees with Camacho that development ethics goes beyond merely the application of existing standard ethical theories (deontologism, consequentialism, and virtue ethics) to problems of development. In his attempt to define development ethics Camacho has taken a closer look at the history of the relationship between technology and ethical ways of life as they have developed unevenly and ambiguously over time. Esquith further agrees with Camacho that agricultural technologies have unintended harmful consequences. They are also in agreement that technologies must be assessed in terms of their social and cultural effects. And so development ethics ought to take notice of harmful social and cultural consequences of new technologies that intensify the production process. Upon further analysis of Camacho's argument, Esquith has established and made explicit some three principles that he believes Camacho commends (although not explicitly stated by Camacho). These are the principle of inclusion, the principle of identity, and the principle of discourse. Esquith has also extended Camacho's argument beyond the subject of proper ethical conduct that moral agents have to the practical issue of collective responsibility. He has demonstrated how we can expand the categories of ethical theory so that they address the issue of responsibility in as nuanced a way as they address issues of inclusion, identity and deliberative discourse.

Chapter 8 by John Otieno Ouko is entitled "Agricultural Intensification: Some Human Rights Issues." Intensification of agriculture has been essential for centuries in meeting the increasing demand of growing world population for food. Over the past few decades, intensification has taken the form of Green Revolution-style farming. However, it seems there are human rights challenges to be faced especially if we look at Green Revolution-style farming in the wider picture of economic globalization. In this paper Ouko investigates whether actions that result from Green Revolution-style farming are consistent with human rights and duties. Whereas food is a basic human right, it seems that attempts to resolve food needs can be constrained by the rights of others. Some of the rights issues associated with the Green Revolution-style farming examined are patent rights, and consumer rights. The chapter also highlights some of the institutional constraints on agricultural intensification and how they may lead to non fulfillment of the right to food among other rights, especially for the poor in developing countries.

Clare Palmer brings some of the leading approaches to environmental ethics to bear on agricultural issues, especially agricultural intensification in Chapter 9. She maintains that one category of environmental ethicists focuses on intrinsic values some of which are: sentience and welfare, life, species, and ecosystems/ecological processes or qualities. If environmental ethics is understood in these ways, then agricultural intensification is significant inasmuch as it either threatens, or alternatively enhances the possibility for, the existence of such intrinsic values. However, environmental ethicists have challenged this approach, arguing for a more pragmatist, contextual understanding of environmental ethics, where policies can be decided on the basis of compatiblism between different approaches. After analyzing the arguments presented by each of the approaches, Palmer concludes that although there are no easy answers to the rights and wrongs of agricultural intensification from within any branch of environmental ethics (whether from those arguing in favor of intrinsic value in nature, or for those who adopt a more pragmatic orientation), it can be hoped that contributions from environmental ethics can help to clarify values at stake in, and arguments about, episodes and instances of agricultural intensification; and that some of the thinking in environmental ethics may contribute to decision making in particular cases.

Lawrence Busch critiques Palmer's contribution to the debate on environmental ethics and agriculture in Chapter 10. Whereas Busch agrees with Palmer that environmental philosophers have devoted little or no attention to agriculture, he is dissatisfied with her arguments concerning agroecosystems, biodiversity, and aesthetics. Busch charges that Palmer discusses the problems associated with valuing ecosystems without going deeper to explain how we can know when we have an ecosystem. Second, whereas agricultural intensification could have deleterious effects on the environment as argued by Palmer, extensification, which accompanies intensification, improves biodiversity. Third, Palmer's argument on aesthetics fails to take cognizance of the fact that what is taken by one generation to be pleasing may not be so considered by the succeeding generations. Moreover, in some places divergent aesthetic and ethical values can be found. Busch has also discussed these important areas in agricultural intensification: shift from supply to demand, standards and intensification, and third party certification.

In an essay entitled "Agricultural Intensification, Environmental Ethics and Sustainability: Some Ethical Observations," Nigel Dower notes that Palmer's chapter focuses on environmental ethics as about intrinsic value of things in nature and the impact of agricultural intensification. Dower is in broad agreement with Palmer and only takes issues with her on some small points of detailed exposition and in terms of the general way she sets up the discussion in terms of her understanding of environmental ethics. Dowers starts by discussing the issue of sustainability which he feels Palmer has not adequately addressed in relation to agricultural intensification. His contention is that the issues Palmer raises are unlikely to be taken seriously by policy-makers unless they are situated in the discourse of sustainability. Here he discusses what sustainable agriculture is, parameters of sustainability, and environmental sustainability as they relate to agricultural intensification. It is within this background that Dower situates Palmer's concerns which are primarily to do with the various ways non-human values come into ethical reckoning and how this impacts on agricultural intensification. These concerns include sentience, life as such, species, and ecosystems.

Chapter 12 by David Fraser began life as an FAO Readings in Ethics Paper. The version published here is revised to fit more directly in line with the theme of intensification. It analyses some of the key developments in the history of animal intensification. Fraser comes to the conclusion that many of the standard claims made by critics of intensive animal production are seriously flawed. These are the three widely held views which Fraser has examined and the conclusions his reasoning has led him to. First, it is being claimed that there is a strong link between intensive animal production and problems of animal welfare. On the contrary, Fraser argues that intensive production systems have some positive effects on animal welfare, and that there is no reason to think that intensive production systems cannot be developed so as to provide high levels of welfare. Second, there is a notion that the development and maintenance of intensive animal production is driven by greed and lack of care by producers. On the contrary Fraser argues that the development of intensive animal production was mainly driven by economic factors beyond the control of individual farmers. Third, there is the idea that the way forward is either vegetarianism or a return to traditional, extensive, small-scale, multipurpose farming. On the contrary Fraser argues that the way forward is to reform rather than abolish intensive animal agriculture.

In Chapter 13 Peter Sandøe focuses his attention on the original FAO Readings in Ethics paper written by Fraser (available from FAO and online at http://www.fao.org/ethics/readings_en.htm). He notes that the paper claims that intensive production systems have some positive effects on animal welfare and that intensive production systems can be developed so as to provide high levels of welfare; that the development of intensive animal production was mainly driven by economic factors beyond the control of individual farmers; and that the way forward is to reform rather than to abolish intensive animal agriculture. Sandøe argues that the original paper does not take a wider definition of animal welfare seriously hence Fraser's narrow approach to the issue of animal welfare, and commends Fraser for broadening the approach in Chapter 12. He continues to question Fraser's rejection of the perspective of those urging greater emphasis on animal welfare/ or animal rights like Tom Regan and Peter Singer. Finally, Sandøe points out some of what he thinks are the flaws and missing links in Fraser's reform agenda.

Richard Bawden draws upon systems theory in Chapter 14 to place Fraser's discussion of intensification and animal welfare into context. He argues that any attempt to both acknowledge and address the integration of values and science will be successful to the extent that it first recognizes the multiple layers of complexity and concern that face livestock producers as the attempt to respond to the pressures of the economic and regulatory (as well as natural) environment. According to Bawden, one of the failures of science-led attempts to promoted development and intensification has been a neglect of the way that they themselves embody a particular systematic approach to knowing that tends to conceal both the value-laden dimensions of science and also the orientations of alternative perspectives. While Bawden commends Fraser for having recognized the complexity of animal welfare within contemporary animal production, he believes that Fraser's approach understates the challenges to be faced in any attempt to address issues of animal welfare within the overarching framework of the ethics of intensification.

Chapter 15, "Ethics in Agricultural Change: Questions and Proposals for Development Processes," by Andrew Dorward notes points of consensus reached at the MSU/Imperial College at Wye Conference on the value and direction for further interactions between philosophers and practitioners of agricultural intensification. He argues that formal ethical analyses of development schemes (not unlike economic analyses currently done) would have the virtue of presenting planners and decision makers with an explicit articulation of the various rationales for undertaking a project on agricultural intensification. Such a tool would be invaluable in helping to identify areas where ethical perspectives converge and diverge in the development planning process. He argues that an iterative approach to ethics would provide roles both for academically trained experts as well as for the voices of affected parties.

Conclusion

In concluding this introductory essay, it is worth stressing that the FAO paper immediately following was itself intended to be introductory to an ethically explicit conversation about agricultural technology and agricultural development. Again, the entire project has been conceived at the outset to be the first, rather than the last, word on the topics covered. In addition to the other FAO papers and documents related to ethics, readers new to this subject are encouraged to consult journals such as *The Journal of Agricultural and Environmental Ethics* or *Agriculture and Human Values* that have been created to serve as a continuing forum for these discussions. And readers are encouraged to join that conversation by adding their own perspectives, reactions and ideas.

Chapter 1
The Ethics of Sustainable Agricultural Intensification

FAO Ethics Series

Agricultural Intensification

Agricultural intensification can be technically defined as an increase in agricultural production per unit of inputs (which may be labour, land, time, fertilizer, seed, feed or cash). For practical purposes, intensification occurs when there is an increase in the total volume of agricultural production that results from a higher productivity of inputs, or agricultural production is maintained while certain inputs are decreased (such as by more effective delivery of smaller amounts of fertilizer, better targeting of plant or animal protection, and mixed or relay cropping on smaller fields). Intensification that takes the form of increased production is most critical when there is a need to expand the food supply, for example during periods of rapid population growth. Intensification that makes more efficient use of inputs may be more critical when environmental problems or social issues are involved. In either case, changes caused by intensification are to be understood conceptually in contrast to extensive adjustments, which involve increases or decreases in the amount of inputs used. Historically, the most common and effective extensive adjustment in agricultural production has been to increase or decrease the area of land planted.

For the purposes of this discussion, the contrast between intensification and extensive adjustment is intended to indicate the contrast between two broad strategies that human beings have had for affecting their food supply, rather than concepts applicable to economic or technical analysis of specific cases. The technical specifications and measurement of intensification or extensive adjustment in any given case are actually quite complex. Changes in the productivity of one input are likely to be accompanied by adjustments in the amount of other inputs. This complexity notwithstanding, there is little doubt that agricultural intensification has been a prerequisite to human civilization. The Neolithic technological revolution was built on collecting, concentrating, selecting and harvesting plant and animal species in an organized fashion, with the aim of having more products closer to hand and easier to convert into nutrition. The domestication of farm animals and the development of crops, in the context of ever more productive farming systems, enabled the human population to grow and towns and villages to develop, having governments, laws, trade and economies with specialized employment.

P.B. Thompson (ed.) *The Ethics of Intensification*,

Applying fertilizer to a maize crop

As agricultural production became more efficient, so populations increased. Historians have argued whether higher populations drove technological development, or whether technological development made higher populations possible. Nevertheless, throughout this development, most societies were chronically malnourished, or prey to episodic famine. Relatively high transport costs meant that most societies relied on local production, except when water transport made imports possible. Classical Athens was largely fed from the Crimea, and Rome from Egypt and southern Spain. For more than 95 percent of the history of civilization, food has been scarce for nearly all people. This has meant low life expectancy, susceptibility to disease and little capacity to face wars, droughts, floods and other human and natural catastrophes. Food scarcity and social disorder brought about major migrations of people and caused wars and massive cultural disruption.

Conversely, when scarcity was relieved, major cultural advances were made. About 1 000 years ago, new varieties of rice were taken to southern China from the Champa Kingdom (now Viet Nam). These were not sensitive to photoperiod, and produced two crops a year instead of only one. When these varieties were grown, the intensive techniques that had developed slowly in China to increase productivity per unit of land, although requiring more labour, resulted in dramatic production gains. The subsequent rice surplus triggered changes across various sectors of China's economy, stimulating the construction of roads, canals, dams, ironworks, grain storage facilities and the production of weapons. For five centuries southern China experienced sustained economic growth and had favourable trade terms for silk, spices and technology with late Mediaeval and early Renaissance Europe.

The agricultural revolution in the early modern United Kingdom greatly increased agricultural productivity. It relied heavily on techniques based on horse power, soil-supporting crop rotations, land drainage and grazing systems, which were developed on the European continent in the very densely populated and often scarcity-ridden Low Countries (the present-day Benelux). When these innovations were applied to larger areas at lower human population densities, a significant surplus was produced, which made the Industrial Revolution possible. At the same time, the harnessing of energy sources became more efficient. The late eighteenth century saw improved water mills and windmills, the use of sea coals and the development of steam engines. Transport costs fell, making it economically viable, for example, to ship bones long distances for use in fertilizer. Internal combustion and hydropower later greatly increased returns to labour and the availability of products such as nitrogen fertilizers.

However, any technology exists within a social and political system. The way that this British agricultural surplus was generated, controlled and distributed under the restrictive Enclosure Acts forced most of the rural population off the land to serve as industrial labour. The result was high food insecurity and a structural form of urban poverty, where families no longer had access to land or to traditional rights such as commons and gleaning. Yet the growing surpluses allowed greater specialization, provided capital and cheap labour for industrialization in Europe and drove the long, violent European nineteenth century. They also resulted in much larger markets in food to supply a burgeoning population, which produced many more goods and services than at any time in previous history. However, the costs paid in human suffering by three or four generations of impoverished families were considerable.

An understanding of the ethical issues involved in intensification can take both a prospective and a retrospective outlook, and ethical standards for evaluating intensification can take either a broad outlook on the general trend of events or a specific focus on the particular responsibilities of key actors. There are three general ethical questions to be posed. First, it is possible to ask whether intensification in a given situation is good or bad, all things considered, without pointing to specific decisions or activities undertaken by particular people or organizations. Second, assuming that intensification is a good thing, how should the burdens and benefits of intensification be distributed? Third, who is responsible for seeing that intensification occurs and that it follows an ethically acceptable path? Beyond these questions, it will be critical to deploy the resources of the natural and social sciences to identify the impediments to intensification, as well as to identify factors that would make an ethically justified form of intensification become ethically problematic, but that task will not be attempted in the present paper.

An Ethics Framework

The word "ethics" refers to principles or standards that define behaviour, action or rules for action that is considered to be right, good and proper. A framework for organizing the enormous variety of ethical standards that have served in this role throughout human history can be developed from a simple schema of human action. Individuals, associations or the designated agents of organizations can each be characterized as *actors*, represented by the shadow person shown in Fig. 1.1. Actors considering or initiating action do so under a set of constraints, represented by the

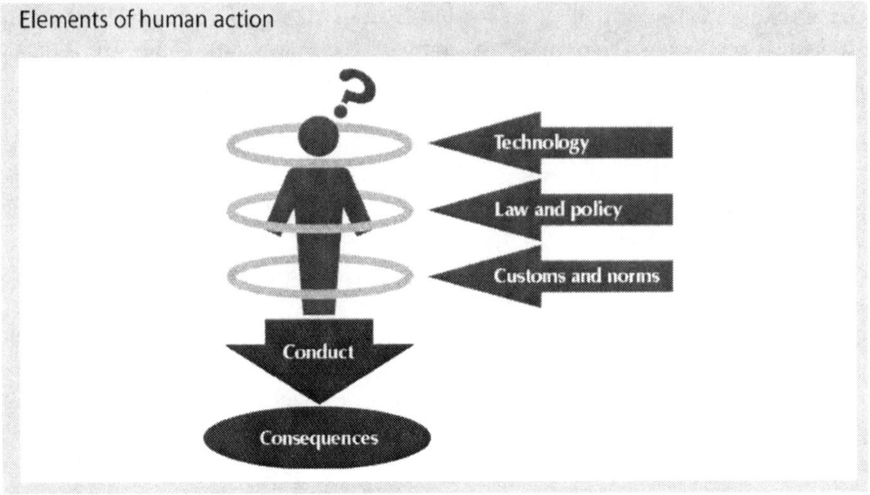

Fig. 1.1 Elements of human action

rings encircling the shadow person. These constraints determine which action or behaviour is possible, and are of three kinds. First, some constraints determine the physical universe of possibility. Characters in science fiction novels may be able to dematerialize and transport themselves to other places, but human beings cannot. Constraints that determine the physical limits of possible actions represent *technology*. Second, *law and policy* limit the universe of possible behaviour and action that an actor will consider. Third, individuals and associations constrain their behaviour according to *customary norms* that often lack any legal or official sanction, yet may function very effectively to limit the universe of possible alternatives for action. For example, people from Western societies will spontaneously queue for service, although in most instances this norm lacks legal reinforcement. Together, these three types of constraint make up the *opportunity set*, the class of actions or behaviours that are effectively available to any potential actor.

Eventually, the actor will select one possible course of action from the opportunity set and will engage in conduct. Conduct indicates the behaviour performed, including both physical motions and symbolic or meaningful behaviour. Conduct may be quite complex, and it is not unusual to characterize a long series of acts or behaviours performed over time as a single action. Because conduct is an active response to an actor's opportunity set, it is indicated by an arrow in Fig. 1.1. Clearly, the significance of behaviour and the interpretation of the relatedness of multiple acts depend heavily on the broader social context. There may be room for differences of opinion about what, exactly, constitutes conduct in a given instance. For the present purposes, conduct is inclusive of all acts understood as components in an actor's performance of an action. One primary goal in offering this definition of conduct is to distinguish conduct from the *consequences* of the agent's behaviour, which can be understood to be the effects of the action on the natural world, particularly on other people and associations, represented by the oval in Fig. 1.1. The term "consequences" here designates especially changes in the health, wealth and well-being of affected parties (including the person who acts) that are caused by the initial action. As with conduct, there may be differences of opinion about what these consequences are, especially when consequences are indirect or are remote in space and time. Notwithstanding these possibilities for difference in interpretation, Fig. 1.1 represents a very simple picture of human action as conduct performed under constraints and producing consequences or outcomes.

Three distinct ways in ethical principles can be developed to determine whether an action is right, good and proper. First, it is possible to see the ethical validity or correctness of an action as a function of its consequences. Increases in the health, wealth and well-being of people are generally characterized as *benefits*, whereas adverse effects on health, wealth and well-being are characterized as *harms* or *costs*. Right, good or proper actions will tend to be seen as those that have achieved the best balance of benefit and harm relative to other possibilities in the actor's opportunity set. Second, it is possible to see the ethical validity or correctness of an action in terms of its consistency with an ideal set of constraints. These constraints may be articulated either as *duties* that the actor must discharge, or as *rights* held by others, which the agent must respect. Rights and duties are generally correlated, however, so that if one person has a right, others have a duty to respect it, while having a duty means that others have a right to expect that the duty will be discharged. Finally, it is possible to see the ethical validity or correctness of an action in terms of conformity to certain types of conduct. Instances or patterns of conduct that are ethically right, good and proper are *virtues*, while those that are wrong, bad or improper are *vices*. This third pattern of ethical evaluation lends itself particularly to expressions of ethical judgement that emphasize the character of the actor, so that not only is the act virtuous, but also the person who reliably acts in virtuous ways.

In summary, a simple analysis of human action indicates three patterns of argument or discourse for articulating, stipulating or defining actions as right, good and proper. Each pattern tends to place the focus or emphasis of ethical inquiry in a different place and many philosophers have developed entire moral systems based entirely on one of these three approaches. In many instances ethical disagreements arise from one party's tendency to formulate a rationale for the evaluation of an action in language and concepts that emphasizes one of the three patterns, while another party emphasizes one or both of the other two. Nevertheless, it is possible for there to be significant differences in approach even within each of the three broad patterns, and many of history's most notable moralists have tended to develop accounts of ethical evaluation that involve considerably detailed discussion of one framework. For simplicity, arguments that interpret the ethics of an action as a function of benefits and harms (or costs) will be called *consequentialist*. Arguments or claims that understand what is right, good and proper as determined by rights or duties will be called *rights-based* and statements that stress the conduct and character of the agent will be called *virtue-based*. It will prove useful to discuss each general approach in slightly more detail while discussing the main topic of intensification.

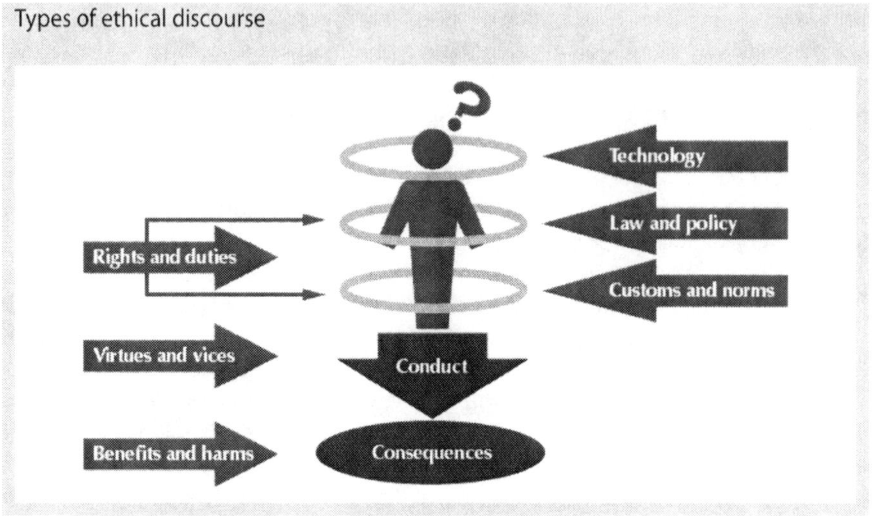

Fig. 1.2 Types of ethical discourse

When is Intensification Ethically Good? A Utilitarian Model

Agricultural intensification is a process that occurs when individual human beings, communities or organizations take actions of one sort or another. The ethics framework previously discussed can be applied to the actions of individuals, associations and organized corporate bodies. The framework illustrates three ways in which questions about the justifiability and ethical acceptability of any particular course of conduct might be posed and analysed. Applied to agricultural intensification, these become questions that might be pertinent to the acts of farmers, input suppliers, technology developers, or to any number of public and private agencies whose activities affect the productivity of agricultural inputs. Such questions might be asked by a person or group evaluating their own options and might also be posed as part of a general discussion and debate about what actions should be taken by governments, international agencies or, indeed, any actor in the food system. Since policies are, in fact, actions taken by governments or other organizations, it is possible to evaluate the institution of any policy that would affect intensification in much the same way as one evaluates any ordinary act. The first ethical question is to determine what it is about intensification that makes it a good thing, something to be encouraged or brought about in a particular set of circumstances and, correlatively, what circumstances might make intensification an ethically bad thing.

Consequentialist ethical approaches provide the most straightforward and obvious way to evaluate an entire system of food and fibre production. The consequentialist understands what is right, good and proper to be determined by the impact of the action or policy on health, wealth and well-being. Intensification is,

Intensification is associated with periods of human population growth

in the prototypical case, intended to increase the total amount of food available without increasing the use of inputs. Since food is material to human life and health, the production of more food can be considered a beneficial impact, especially under the circumstances of food scarcity that have been too typical of human history. As noted, intensification is associated with periods of human population growth. Without a correlative growth in food supplies, food scarcity causes hunger, disease and starvation. Using the framework previously described, many individuals and groups consider options and undertake conduct that has the outcome (consequence) of increased food supply. The benefits associated with increased food availability provide the elemental argument for intensification, and this argument is consequentialist in its moral logic. In the simple case where new technology or farming methods allow a farmer or landowner to produce more food, consequential reasoning shows why this is ethically a good thing.

As already stated, European agricultural intensification immediately prior to the Industrial Revolution was accomplished not only by applying a package of new production technologies to farming, but also by the Enclosure Acts, which disestablished a system of rights and duties that permitted commoners to live on and farm lands as long as their crops were shared according to an ancient formula. The framework applies not only to the conduct of individual farmers and landowners, but also to the political activity that led to this policy change. Was enclosure ethically justifiable? The British philosopher John Locke (1632–1704) posited the following argument to show that it was:

> … He that encloses land, and has a greater plenty of the conveniences of life from ten acres, than he could have from an hundred left to nature, may truly be said to give ninety acres to mankind; for his labour now supplies him with provisions out of ten acres, which were but the product of an hundred lying in common.[1]

Here the disestablishment of the old system of commoner rights and duties is portrayed as justified in light of the increased benefits (conveniences) accruing from enclosure. Although this style of thinking is not typical of Locke, the passage implies that any system of rights and privileges is justified, given the efficiency with which it supplies human beings with provisions.

Efficiency is particularly important in the most common form of consequentialism, *utilitarianism*. Utilitarians assume that the values associated with consequences can be quantified to produce a ranking system for all possible courses of action (or *options*) available to an agent. They also assume that the value of benefits and harms can be added and subtracted. Such a ranking system produces a class of optima such that no option in the opportunity set yields greater total value (although there may be more than one option that is optimal). According to the utilitarian standard (e.g. the *utilitarian maxim*), the right, best and proper action or policy must be a member of this class of optima. This is popularly stated as: Act so as to produce the greatest good for the greatest number of people. Thus, the most efficient approach to producing benefits or avoiding harms is the course of action most thoroughly justified by ethics. Utilitarianism has been an implicit ethical philosophy for agricultural science, which has sought to "make two blades of grass grow where one grew before".

One critical and often overlooked aspect of the utilitarian approach is the need for a complete accounting of costs and benefits. The green revolution involved new seed varieties that were more responsive to nitrogen fertilizers, which are, in most settings, a purchased input. Hence, a simple utilitarian approach weighs the benefits of increased yields against the costs of seeds and fertilizer. If benefits outweigh costs, the green revolution is justified. Yet other shifts accompanied the new technology, and many critiques of the green revolution can be articulated entirely within the framework of a utilitarian/consequentialist ethic. Within the first decade of the green revolution of the 1970s, which led to aggregate increases in rice production across Asia, large-scale insect pest outbreaks and plant disease epidemics destabilized food production, supplies and prices.

Only after three boom and bust cycles and pest outbreaks did governments begin to move away from simplified, centralized pest control policies that relied on insecticides

---

[1] J. Locke. 1690. *Second treatise of Government*. In C.B. McPherson, ed. 1980. Indianapolis, USA, Hackett Publishing. Locke's philosophy is not consequentialist, but contractarian. He believed that people had a natural right to appropriate goods (including land) found in nature, and that others had a duty to respect this property right, which was grounded both in the nature of things and in the social contract forming the basis of civil society. It is thus likely that he understood the phrase "give ninety acres to mankind" in an almost literal sense, and saw the justification of enclosure in terms of a kind of expansion of the commons, rather than in starkly utilitarian terms. Nevertheless, it is difficult to interpret this particular passage as anything more than a consequentialist moral argument.

and vertical host plant resistance towards decentralized integrated pest management that built on local ecological processes to realize production potential. With the concentration of farm animal processing facilities, the chances of large-scale epizootics occurring (such as foot-and-mouth disease) increase exponentially because of the more extensive movements of animals between pasturage, feedlots and abattoirs, and contacts with animal offal and excrement. Feeding livestock with products derived from their own species creates routes for infection by diseases associated with prions.[2] Short-rotation forest plantations increase wood (especially pulpwood) production but at the same time increase vulnerability to specialized pests and diseases. Fast-growing dwarf coconut varieties increase short-term yields but are more frequently at risk from diseases previously found in limited geographic areas. In fisheries and aquatic production systems, exotic species are commonly introduced. These often initially increase total production, but can unexpectedly change trophic relationships and disrupt ecosystems, as did the Nile perch in Lake Victoria. Intensive salmon hatcheries have been criticized for reducing the genetic adaptability of natural populations. These problems testify to the need for completeness in thinking through the costs and benefits of intensification.

When is Intensification Ethically Good? A Rights-Based Model

John Locke's rationale for enclosure makes no mention of rights, but the language of rights might be used to develop a rationale for arguing that efficiency-seeking public policies (such as enclosure) are ethically wrong in certain circumstances. Simply stated, policies are unjustified if the changes they endorse violate or override important rights. Among the rights that might be violated are property or occupancy rights, if enclosure forces people who hold such rights off the land, or subsistence rights, if the effect of enclosure is to put some people in a position where their rights to food and shelter are not met. Yet the ethical principles behind these rights are complex.

A rights-based approach to ethics (sometimes referred to as *deontology*) proceeds by stipulating or deriving a set of basic rights and duties that agents must perform without regard to the consequences that might arise in any particular case. A rights-based approach to intensification is more concerned as to whether the actions that result in higher food production are consistent with these rights and duties than in their eventual effect on human welfare. Several methods have been put forward for identifying these rights and duties. The German philosopher Immanuel Kant (1724–1804) developed complex philosophical arguments for a master principle he called the *categorical imperative*: never act in a way that treats

[2] Prion-associated diseases include kuru, scrapie and bovine spongiform encephalopathy (BSE). The infection route and aetiology of BSE and the variant of Creutzfeldt-Jakob disease that has been linked to BSE are still a matter of investigation.

another person solely as a means to an end. For a Kantian, rights and duties can be traced back to this principle, which tells us always to respect other people's capacity for freely choosing their life plans. The problem that a Kantian would have with Locke's argument for enclosure is that it seems to treat the commoners (whose rights are violated) simply as a means to the larger end of increasing food production.

Kant's deontology came at the end of a longer tradition of *natural law theory*, which held that the basic rights and duties forming the main content of the (ideal) moral law are evident to any rational person, and have hence invested great effort into rational argument for certain approaches to the configuration of rights. Kant argued that every human being wants to be treated as a free, autonomous agent, and that consistency requires people to treat others in the same way. For other philosophers, going back to Thomas Hobbes (1588–1679), rationality was understood simply as enlightened self-interest. The purpose of rights is to protect people's ability to act on their own enlightened self-interest. Hobbes believed that any rational person would agree to live in accordance with certain rights and duties, if only they could have a reliable expectation that everyone else would do the same. This approach is sometimes characterized as *contractualist* in light of the way that rights and duties are described as being grounded in an implicit agreement (or social contract) among all members of society.[3]

While a Kantian might argue that enclosure treated commoners as a mere means to a larger social goal, a contractualist might say that enclosure violated the social contract. In either case, the problem with enclosure from a rights-based perspective was that commoners had a right to use these lands. As such, any plan to exclude them from the land could not be justified unless it included some provision for obtaining their agreement. This would require involvement of the commoners at some stage of the enclosure process, and their involvement would need to be such that each rights holder had the opportunity to give or withhold their agreement voluntarily to the plan. Perhaps they could be convinced that they would be better off, or perhaps they would be enticed to accede to such use of the land in exchange for compensation. These details might vary considerably on a case-by-case basis, but what is critical from a rights-based perspective is that respect for the individuals affected by enclosure requires that they be accorded a role in the intensification process that is fully consistent with their rights. Either version of rights theory provides a starting-point for questioning whether intensification is good simply because it produces more benefits (in the form of greater food production) than costs (in the form of losses for the minority).

---

[3] Although similar in important respects, Kantian deontology and Hobbesian contractualism provide important different rationales for justifying rights. In both versions of rights theory, the ethical significance of a right resides in the way that it protects human freedom. For Kantians, an individual's freedom is an expression of the ability to plan and order one's own thoughts and actions rationally, while for contractarians freedom means simply that others do not control or limit one's action. In Kantian philosophies, preserving the entitlements owed under the system of rights is a form of showing respect for another's need to plan and order his or her own life, but rights can also be seen as being based on the social contract.

Either approach to rights implies that the exclusion of commoners from decision-making is unacceptable, but a Kantian might in addition note that the effect of enclosure was radically to impoverish commoners and their descendants. For a Kantian, such conditions of poverty make it impossible for a person to exercise rational free will; the circumstances of need are so great that people in dire poverty are effectively coerced into enduring humiliation and deprivation. As such, key subsistence rights, including the right to food, become minimal conditions that must be met if all people are to be treated with the moral respect to which they are due. Thus, any situation in which people are so poor that they cannot freely exercise their innately human capacity to choose a life plan involves an ethical wrong. Methods of intensification that place people in such circumstances cannot be endorsed from a rights-based perspective.

Utilitarianism and Rights-Based Ethics: Further Issues

The basic tension between utilitarian consequentialism, on the one hand, and rights-based ethics, on the other, underlies many issues associated with agricultural intensification. For decades, researchers from developed countries have harvested germplasm from farmers and local markets in the developing world. These researchers have used the germplasm in breeding programmes to develop higher-yielding varieties as well as searching for other valuable genetic traits. Many of those who collected seeds were shocked when critics suggested that their work

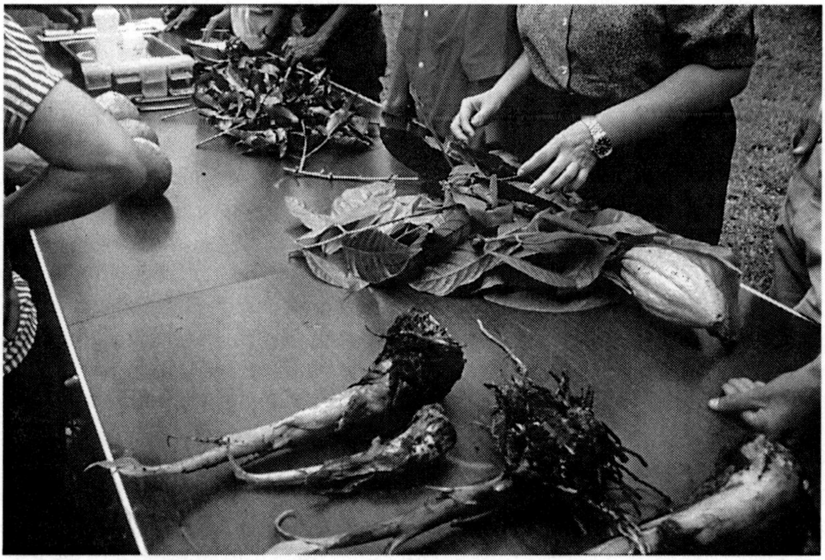

Local scientists at a training session on the collection of *germplasm* in vitro

failed to respect the rights of people from the developing world. From a utilitarian viewpoint, the increased yields of new varieties more than justified the collection of germplasm, and researchers saw no ethical issue in using seeds they had collected this way. However, critics asserted that researchers had failed to show proper respect for the rights of indigenous farmers whose forebears had saved seed for centuries. Purchasing seed in village markets, critics affirm, gives the buyer an implied right to use the commodity good for food, or possibly for replanting, but farmers could not be interpreted to have given up rights to further development of their germplasm without a careful and explicit process to inform them of its true value and to ensure that they had given consent. Some critics argue that because of the collective and collaborative nature of seed development in traditional agriculture, only someone who represents the collective interests of all growers would be in a position to undertake such a negotiation.

Today, some opponents of genetically engineered crops argue that individual consumers should not be forced to eat these crops against their will. Advocates of genetically engineered crops see them as a safe and effective tool for increasing the efficiency of farm production and believe that their opponents' claims are an unjustified barrier to adopting them. While this debate often involves factual disputes about the safety of these crops, the underlying ethical structure of the debate pits the rights-based claims of the opponents against the utilitarian reasoning of the advocates. Why? Irrespective of the safety or risks associated with genetically modified (GM) crops, opponents are claiming that the companies promoting these technologies have placed food consumers in a position where they have no opportunity to reject them. Rather than rebut the claim that consumer rights are at stake on its merits, advocates have often argued that if the crops are safe (as they claim), then consumers have no basis for rejecting them, since rejecting GM crops is an action with real costs but no real benefit. Thus, opponents try to meet a rights claim with a utilitarian argument, and the ethical issues fail to be enjoined.

A similar point of tension arises in debates over the green revolution technologies. When higher yields are associated with purchased inputs, those with access to capital will have an advantage over those without. One can interpret losses experienced by poorer farmers as additional costs and weigh these, too, against benefits from higher yields. The study *New seeds for poor people* by Michael Lipton and Richard Longhurst (1989) is a particularly exhaustive and theoretically sophisticated attempt to assess the green revolution varieties from a utilitarian standpoint. The authors conclude that, over time, these technologies have been, on balance, beneficial to poor people. However, these arguments do not necessarily address concerns that rights may have been violated, or that cultural traditions may have been lost as a result of green revolution strategies. How might rights be affected? It is possible to see any transition from a situation in which people can feed themselves and meet their needs to one in which they cannot as a violation of their rights. Even if such transitions have benefits that outweigh the cost, they would not be seen as justified if there are even a few individuals whose rights to subsistence are jeopardized as a result of the changes.

This is only a cursory introduction to the way that rights arguments should figure in an overall evaluation of the green revolution. Yet one of the problems that has arisen in debates over the retrospective impact of green revolution varieties is that those who draw their ethical norms from utilitarian thinking seem to be ignoring ethical claims that draw upon the language of rights. In doing this, people create an impression of, at best, insensitivity to the full range of ethical concerns relevant to intensification (which should also include virtues, discussed later) and, at worst, arrogant dismissal of arguments that are inconvenient for the case they wish to present. When people who hold influential positions for future attempts to meet food needs dismiss alternative arguments in this way, they engage in a use of power that is itself ethically questionable. As such, it would be valuable to launch retrospective studies that make an explicit attempt to acknowledge the full range of concerns that have been or might be brought to bear on evaluating green revolution intensification, in part as preparation for more open and informed debates on the questions that must be addressed if world food needs are to be met in the twenty-first century.

These cases are complex and deserve a more careful analysis than these summary statements can provide. The point here is simply to note that, in both cases, a utilitarian/consequentialist rationale is met with counterclaims that assert rights. Someone assuming that utilitarian models are appropriate will respond to these assertions with further arguments reciting the costs and benefits of alternative arrangements, but to the extent that these rights are thought to be moral rights, rights protecting the dignity of involved parties, recitation of further costs and benefits will simply miss the point. What is being claimed is that there is a need to respect affected parties by involving them fully and non-coercively in the process of intensification. From the perspective of rights-based ethics no recitation of costs and benefits will justify a failure to do this; what is needed is a justification that addresses how the process of intensification influences the freedom of all persons affected by it. If freedom is being constrained, it must be shown that this constraint is justified, perhaps because it is required in order to show proper respect for the rights of less fortunate people, or perhaps because the people agree to the constraint of their own free will.

When is Intensification Ethically Good? A Virtue-Based Model

The utilitarian/consequentialist and rights- based approaches are both theoretically well-developed articulations of ideas used every day by people who are attempting to determine the correct or proper course of action. However, it may be more typical for people to associate ethics with less systematic ways of thinking. For example, many people may address ethical questions by asking themselves how some exemplary person would act in a given situation. This exemplary person might be a family member, a revered and respected member of the community, a religious leader or perhaps even a person from legend or history who is not even known as a real flesh-and-blood individual. In any case, one draws upon a mental image of how a good person would act in the situation at hand. By doing this, one understands the ethics of the situation in terms that refer directly to the conduct that is being performed (see Fig. 1.2), rather than either the rights-based constraints or the consequences of the action.

The philosophical possibilities for developing this general approach to ethics are numerous. The Greek philosopher Aristotle (384–322 BC) did so by developing a catalogue of both positive and negative exemplary types of conduct, or virtues and vices, respectively. He theorized that all human tendencies had appropriate forms of expression in moderation but could become vicious if not held in check. A good person is one who strikes the mean position among these tendencies. Moderation (or temperance) itself thus became the sovereign virtue for Aristotle and many of his followers. Today, many types of ethical theory that emphasize conduct, rather than rights and duties, on the one hand, or consequences, on the other, are referred to as Aristotelean or, more accurately, virtue theories. Although some philosophers have believed that virtues and community traditions are relatively undeveloped sources of ethical insight, better articulated by emphasizing rights and duties on the one hand, or consequences on the other, others have argued that this domain of ethical ideas or source of ethical insight cannot be eliminated.

Aristotle also believed that one's tendencies (as well as one's abilities to regulate them) are a reflection of the sociocultural environment in which one lives and is raised. Thus an exemplary Greek might have a moral character quite different from that of someone who had not lived in what was, in Aristotle's day, a culture almost unique in its egalitarianism, its emphasis on education and its ideals of citizenship. One contemporary school of ethics that stresses the social roots of exemplary conduct is called *communitarianism*. Here, the articulation of ethical norms and standards is likely to call attention to the norms, practices, traditions and institutions that are particularly characteristic of and valued by a particular community, rather than particular virtues and vices. Like Aristotle, contemporary communitarians emphasize the need to have a social environment, a form of community life, that will give rise to exemplary conduct and that will allow people to appreciate the ways of life that such forms exemplify.

Agricultural ways of life have figured prominently in some of the most influential articulations of virtue and vice. The ancient Greeks themselves developed a form of agriculture based upon rough terrain, varying soil types and a Mediterranean climate. Their farms were a diverse mixture of grain production and pastoral livestock, but with heavy reliance on tree and vine crops. The mix of crops and long growing season provided steady work for fairly small households all year round, while the trees and vines involved lifetime investments for smallholders. Military historian Victor Davis Hanson argues that this pattern of agriculture gave rise to unique forms of military organization and tactics as well as the political culture of the city-state. The relatively large proportion of the population controlling property and the nature of their stake in the land made them both fierce defenders of egalitarian political forms and equally fierce warriors who could be relied upon for phalanx manoeuvres requiring discipline and loyalty. These character traits, so critical to the success of Greek city-states as political and military entities, were thought to emerge naturally in a farming population of smallholders. In contrast, the large-scale plantation-style irrigated agriculture common among the Greeks' military rivals relied on stratified societies of slaves and masters who did not develop the requisite virtues.

The idea that forms of agriculture were seminal sources for community practice and national culture reached its culmination in the intellectual cultures of seventeenth and eighteenth century Europe. These ideas were especially influential for those who framed the United States Declaration of Independence and Constitution. In his *Notes on the State of Virginia* (1781–84), Thomas Jefferson, third President of the United States famously wrote:

> Those who labor in the earth are the chosen people of God, if ever he had a chosen people, whose breasts he has made his peculiar deposit for substantial and genuine virtue. ... The mobs of great cities add just so much to the support of pure government, as sores do to the strength of the human body.

Jefferson's key idea was that smallholding farmers would have a greater stake in the stability and success of the new nation than either manufacturers or their labourers, since both of the latter could pull up stakes and leave when difficulties arose. As President, Jefferson went on to set a course for the development of the United States as an agrarian nation, negotiating the Louisiana Purchase to ensure ample lands for future generations of American farmers and authorizing the (Meriwether) Lewis and (William) Clark expedition to determine the suitability of these lands for cultivation and transport.

Jefferson's plan was not, of course, an episode of intensification, as his strategy called for extensive expansion of American farming rather than a transformation designed to increase yields or use resource inputs more efficiently. Nevertheless, his view is important because it shows that ideas about how farming systems and methods produce virtues such as citizenship and community solidarity have had a profound influence on political developments in the past. Plans for intensification that substantially alter the pattern of land tenure, or that change the basic practices of farming thought to be critical to the formation of exemplary patterns of conduct or community identity are almost certain to provoke moral protest. Indeed, the most memorable protests against British attempts at enclosure are not tracts arguing that rights have been violated, but literary efforts such as Oliver Goldsmith's poem, *The deserted village*, lamenting the loss of small village cultures thought to be particularly characteristic of the British national character:

> Ill fares the land, to hast'ning ills a prey,
> Where wealth accumulates, and men decay:
> Princes and lords may flourish, or may fade;
> A breath can make them, as a breath has made:
> But a bold peasantry, their country's pride,
> When once destroy'd, can never be supplied.[4]

The Old Order Amish communities established throughout Europe and now dispersed around the globe provide another example of a philosophy of agriculture

[4] Oliver Goldsmith. 1770. *The deserted village*. Accessed online at: www.ucc.ie/celt/online/E750001-001/.

Intensification: implications for reservoirs and fisheries

The twentieth and twenty-first centuries have seen an unprecedented increase in the creation of reservoirs throughout the world. The two major purposes of dams have been the production of hydroelectric power and storage of water for irrigation. Nevertheless, damming a river to create a reservoir has a major impact on water flow, the aquatic environment and life. From a utilitarian perspective, the costs associated with these impacts (as well as the actual direct costs of building a dam) are justified where the benefits from hydroelectric power and irrigation are greater. Arriving at the optimal relationship between costs and benefits generally involves compensatory measures that mitigate the costs associated with the impact on fisheries.

To the extent that hydroelectric power and irrigation water address problems of poverty and need, there can be a strong rights-based argument for building dams. Yet reservoirs often involve the most dramatic and irreversible transformations imaginable. From a rights perspective, the key issue is that an effort to address human rights to basic energy and food needs may come into conflict with existing land use and riparian rights. Rights arguments might be raised against dam construction when it seems that affected parties either have not or cannot be feasibly brought into decision-making. Others might argue that important traditions, communities and ways of life are lost both when dams involve the flooding of native homelands for large communities, and when traditional fishing methods can no longer be practised. These latter concerns appeal more to the ethics of virtue and vice; they suggest that what is important is being able to live in a traditional way and to engage in traditional practices. But dangers to fish stocks can come from water pollution and from overexploitation by fishing communities themselves. Here, an existing or traditional pattern of rights may be permitting conduct that has significant adverse consequences.

An additional problem is that the distribution of benefits from intensification, hydroelectric generation or irrigation may not reach those rural communities most affected by the increased development. Fishers are usually from a poor sector of society and are often ignored in decision-making.

that relies heavily on virtue ethics. The Amish emphasize a very high degree of family and community integration together with independence from the outside world. They derive meaning from their ability to live together at a particular place with a great stability of practice over generations. Agriculture is important because Amish communities see dependence on outsiders as a potential threat to their ties to one another. The Amish are notoriously suspicious of modern technologies, largely because they see their effects as weakening social relations among members of the local community. Nevertheless, Amish farmers are known for both high yields and ecologically sustainable farming methods. From an Amish perspective, intensification would not emerge as an ethically important goal, and intensifying practices that weakened community bonds – either by tempting members away from the household or by increasing dependence on the outside world – would be resisted. However, intensification would not be seen as an evil in itself, and increases in yields that could be attained by more effective use of household labour might be deemed entirely acceptable.

These few examples illustrate how thinking in terms of virtues provides an entirely different point of view from which to evaluate intensification. From the perspective of a virtue theorist, agriculture's role in forming both personal and national character provides the basis for evaluating policies and technologies that transform the food system. Periods of intensification would be justified only if they reinforce this role and would be opposed if they tended to weaken it. The actual forms that a virtue perspective might take will be highly variable and will depend upon cultural traditions and history. Thus, while utilitarian/consequentialist and rights-based approaches in ethics point towards ethical standards that might be applied to virtually any agricultural system, the specific content of a virtue approach is likely to be highly dependent on local culture and may vary from one cultural setting to another.

How Should Burdens and Benefits be Distributed?

Utilitarian evaluation, as described previously, is notoriously insensitive to the distribution of benefits and harms across society. As a result, the question as to how burdens and benefits are distributed is raised in acknowledgement of the widespread

Intensive livestock farming provides food for growing populations but raises food safety, equity and animal welfare issues

Intensification: implications for livestock

Intensification in the livestock sector could produce more food for growing populations, but there are ethical issues relating to livestock's resource use, food safety and quality, equity and animal welfare. Pollution of land, water and air from intensive livestock production and processing in both developed and developing countries has become a widespread phenomenon, often acting as a vehicle for disease transmission. Moreover, there are direct issues associated with the transmission of diseases and general food safety. Overconsumption of animal products also carries a number of human health risks. While an increase in the consumption of animal products in developing countries would be highly desirable in combating malnutrition, it would not necessarily be wise for these countries to follow the dietary practices of wealthy nations. Livestock consume about one-third of total grain production, with associated pressure on land and other natural resources, and use of fossil fuel. It is possible to ask whether this grain should be fed to animals rather than people. These environmental and food safety risks represent a catalogue of the possible costs that must be weighed against the benefits associated with the intensification of livestock production. However, it is also possible to address these questions in terms of rights and virtues. Do people have a right to eat what they want even if their diet is found to be risky? Do people have rights to be protected from environmental and food safety risks and, if so, should this protection stress constraints on producer behaviour, or informing and educating consumers as to the nature of the risks? Do traditional dietary practices thought central to people's cultural identity ever override the risk/benefit comparisons conducted in a utilitarian manner?

There are, furthermore, ethical issues about the distribution of benefits. While growth in demand for animal products seems to offer opportunities for the rural poor, to date the large majority of these rural people have not been able to take advantage of such opportunities. Thus, there are important ethical issues about the fairness of efforts to intensify livestock production, and the effects that such efforts have on traditional rural communities. Intensification in the livestock industry also involves a set of issues that concern animals themselves. If unregulated, intensification of livestock production is associated with animal management practices that do not allow the expression of natural behaviour. Should these issues be addressed in terms of welfare tradeoffs between human beings and animals, or is it plausible to argue, as some have, that duties and constraints on human behaviour should be recognized as amounting to animal rights?

feeling that improvements in efficiency and general welfare can come about in a very unfair manner. One possible response is simply to argue that, in questioning whether intensification is fair, we are calling for the methods and processes to be evaluated in terms of their consistency with a concern for human freedom (e.g. an appeal to human rights), or with respect to their impact on traditions and community integrity (e.g. an appeal to virtues). Thus, one might say that utilitarian thinking provides a basis for saying why, other factors being equal, intensification is a good thing, while rights and virtue approaches sensitize us to the other issues that must be attended to in order for intensification to be fully justified.

However, it is also possible to address the question in more classically consequentialist terms. Doing so requires one to develop and defend criteria that can be applied to the way that costs (or burdens) and benefits that are the outcome (or consequence) of intensification are distributed among those affected. The standard utilitarian view suggests that distribution is not important because it is the net or average impact that matters. Yet one could argue that only outcomes in which no one is harmed are ethically acceptable. Another possibility is to minimize the chance of the worst possible outcome, an approach that may reflect the implicit decision strategy of poor societies trying to fend off the risk of total starvation. Another view, adapted from the philosopher John Rawls (1921–2002), is to recommend the option that has the greatest expected value[5] for the poorest group within society. This approach allows one to develop ethical justifications that favour the interests of poor or marginalized people over those who are better off.

Although rights theory is sometimes offered as an answer to questions of distributive justice, difficulties also arise within the rights-based approach. In particular, rights-based thinking is occasionally confronted with situations where rights seem to conflict. For example, the Universal Declaration of Human Rights notes that all persons have a right to food. If one encounters a situation in which the only way to secure this right is to violate other rights, such as property rights, which rights have priority? As in the case of establishing the validity of rights, resolving conflicts among rights will require use of philosophical arguments that support rights-based views. Henry Shue's *Basic rights* argues that rights have an internal principle of order that can be used to resolve conflict. Some rights (such as the right to education or to vote) would not be meaningful unless more basic rights (such as the right to food and personal security) were already secure. Shue argues that basic rights should be secured for all before less basic rights are secured for a few. Shue's approach does not rely on cost-benefit thinking as a "tie-breaker", yet few would resist combining principles from utilitarian/ consequentialist thinking with approaches from rights theory.[6]

[5] Consequentialism is frequently used in connection with analytical techniques that allow one to assign a probability to several different possible distributions of benefit and harm that might ensue after the selection of a given act or policy alternative; thus, an expected value can be prospectively associated with each act or policy in the opportunity set.

[6] Indeed, one of the most sophisticated versions of utilitarian thinking (R.M. Hare's two-level utilitarianism), argues that the traditions of rights provide important sources of moral insight; only when we are very sure that the consequences of our actions are fairly narrow and can be predicted accurately should cost-benefit considerations be allowed to override a traditional rights-based claim. (See Henry Shue. 1980. *Basic rights*. Princeton University Press. 2nd ed. 1996.)

Virtue-based approaches typically address distributive issues either through the idea of community – so that community solidarity comes into play when a few individuals may be put in a position where unreasonable burdens are placed upon them – or as components of a specific virtue, such as charity. In some settings, culturally sophisticated mechanisms for sharing burdens can be imbedded in social norms that would be articulated in the standard terminology of cultural identity, community and personal virtue. However, some articulations of virtue have been particularly insensitive to social inequalities. Virtue-based thinking (often with religious backing) can be used to rationalize enormous inequalities in defense of a given social order. It is no accident that the word "aristocracy" derives from "Aristotle".

Who is Responsible for Ensuring that Intensification Occurs?

The 1996 World Food Summit recognized that the then world population of 5.8 thousand million had 15 percent more food per capita than the global population of 4 thousand million only 20 years earlier. The productive potential of technological improvement was demonstrated in the green revolution of the 1960s, where improved varieties resulting from scientific breeding, largely in the international public sector, coupled with investment in irrigation, mineral fertilizers and institutional infrastructures, raised food production and productivity, particularly in rice and wheat in Asia. But the uneven success of green revolution technologies, with little uptake in Africa particularly, and its unforeseen social consequences, showed that sustainable intensification was not just a result of technological development but of government policy, and social and economic forces. In the last decades, falling transport costs, together with the development of global markets and growing trade liberalization, have given new focuses to the challenge of ensuring a steady reduction in poverty through securing and maintaining adequate long-term production levels as well as conditions of adequate access to food. There are major new factors to consider, including changes to the biosphere resulting from global warming, the as yet unproven potential of the new biotechnologies, unprecedented urbanization and the reality of a world economy without a global economy or global society.

Utilitarian/consequentialist and rights-based approaches in ethics have tended to portray responsibility in very general terms: *everyone* has the responsibility for acting in ways that produce the greatest good for the greatest number, or in accord with duties to respect other people's rights. Thus, if intensification is considered to be a good thing, everyone is responsible for doing whatever he or she can to make sure that it is achieved. Australian philosopher Peter Singer has published a number of essays in which he puts forward this argument exactly, at least with respect to responsibility for ending hunger. However, this view has been the subject of some criticism. It seems to impose an overwhelming responsibility on ordinary people, as if one cannot divert one's time and resources to enjoy life while someone else is hungry. Furthermore, it creates a situation in which nothing is actually done: if something is everyone's responsibility in general, it tends to be seen as no one's responsibility in particular. Singer has acknowledged that this may in fact be a weakness of the utilitarian philosophy that he has advocated.

Intensification: implications for forestry

At present, most wood production comes from natural and semi-natural forests, which cover 95 percent of the world's forest area. Human interventions in forests may aim at very different objectives, not exclusively targeting increases in industrial wood production, but also non-wood forest products and the provision of conservation, protection, recreational and other environmental and social functions. Total demand for wood products is increasing globally and the yield productivity of forest plantations is higher than that of natural forests. However, a major ethical question regarding the intensification of forests is whether it is acceptable to clear natural forests to establish forest plantations. According to current estimates, about 50 percent of newly established forest plantations relied upon the clearing of natural forests, particularly degraded and/or secondary natural forests. These decisions generally stress criteria of optimization typical of utilitarian ethical decision-making.

In the case of natural forests, intensification is not a common concept or priority. Some of the reasons for this relate to ethical concerns. Natural forests are managed for a multitude of functions to provide a large range of products and services (non-intervention, like absolute protection of a conservation area, is a management option). This multipurpose strategy limits options for intensification and specialization on largely utilitarian grounds, but different groups of people have different expectations of natural forests, frequently seeking to harvest different products of the forest or trying to obtain different services from it. The equation "specialization + intensification" often creates conflicts between these groups. Such conflicts raise questions about who participates in decision-making and when formal criteria for involving affected parties need to be specified. Furthermore, in natural and semi-natural forests, management cycles are much longer than in agriculture and require a series of step-by-step, long-term practices. The purpose of forest management can change markedly over time. Changes in the way that forests contribute to lifestyles, community and cultural identity and the broad patterns of social organization are difficult to articulate in the traditional utilitarian language of costs and benefits.

Ethical ideas that stipulate particular responsibilities for people who have special roles – such as teachers, holders of public office, technical experts and parents – have more typically been articulated in the language of virtue and community solidarity. Thus, community leaders are people who have assumed or been appointed to a particular social station that entails special duties. A virtuous leader therefore assumes duties that are not those of the ordinary person, and in traditional societies such leaders would also have the authority to ensure that these duties are carried out. Consequently, in a traditional society, leaders might be acting rightly even when they order people to do things that might not be consistent with a modern conception of rights. Military leaders, for example, can order citizens to sacrifice their lives for the greater good of the community, and one can easily imagine situations in which changes in land use or farming practice might be ordered as part of a leader's performance of special responsibilities.

In modern societies, social roles are often highly rationalized so that particular agencies are formed to take over roles that might have been understood as components of a leader's virtue in traditional societies. Thus organizations such as the Food and Agriculture Organization of the United Nations (FAO) have been assigned an explicit social mission that entails a responsibility to ensure that food needs are met. The report of the 1996 World Food Summit makes the case for a new round of intensification, and places responsibilities on national governments, with organizations such as FAO having major responsibilities for coordinating and facilitating that work. However, officers within FAO have an ethical challenge in meeting this responsibility. On the one hand, traditional models for interpreting the responsibilities these officers have been delegated to carry out stipulate ethics of virtue, and many expect these people to act as virtuous and authoritative leaders, especially when working in traditional societies. These traditional ways of understanding ethical responsibilities provide an implicit basis for acting ethically to discharge official responsibilities. On the other hand, the rationale for forming organizations such as the United Nations has largely been articulated in utilitarian or rights-based terms, and this language places significant constraints on the authority of officers who occupy posts within these organizations. Thus, the exercise of these role responsibilities requires ethical resources that may exceed those on which the mandate of an organization's authority is based.

Conclusion

Population growth provides the basis for a utilitarian argument for intensification, and traditional virtues of leadership might provide an ethic for pursuing intensification for the good of society. In modern societies, the authority to pursue such good

is constrained by the rights of others. It is critical, therefore, to articulate the ethical rationale for undertaking projects of intensification in terms that draw upon each of the traditions available for specifying and critically evaluating an ethical responsibility. Omitting any one of these ways of framing ethical issues results in a weakened capacity to articulate, debate and ultimately assume ethical responsibilities that may arise in connection with population growth and the attendant imperatives for intensification.

Chapter 2
Doing Ethics in Food and Agriculture

Clive Stannard*

Introduction

In recent years, increasing attention has been paid by national authorities in many countries, by civil society and the general public and by intergovernmental organizations to a variety of ethical questions of relevance to food and agriculture. Major recent changes in the fields of food and agriculture – including accelerating technological development, changes in the resource base, and economic and market developments – have also brought to the fore a number of ethical questions of relevance to food security and sustainable rural development. The technical questions that arise within these fields are frequently very specific to these crucial and complex fields of human endeavour, upon which world food security now and in the future depends. It is therefore clear that ethical reflection upon questions of food and agriculture must be grounded on a clear understanding and appreciation of the specificities of these sectors.

The Food and Agriculture Organization of the United Nations (FAO) is the international apex organization dealing with all aspects of food and agriculture, including forestry and fisheries. FAO has therefore taken a number of steps to approach the application of ethical perspectives to food and agriculture in a more systematic way, which I shall first describe below.

I shall then discuss some of the questions with which we have been grappling in developing the FAO ethics programme, and the thinking behind the paper on "*Agricultural intensification: an introduction to the ethical issues*", which was the subject of our discussions at the Conference on the Ethics of Sustainable Intensification, hosted by Imperial College, London, at its Wye Campus.

*Senior Liaison Officer, FAO Commission of Genetic Resources for Food and Agriculture, retired

P.B. Thompson (ed.) *The Ethics of Intensification*,

Development of the FAO Programme on Ethics in Food and Agriculture

Coordination FAO'S Activities on Ethics in Food and Agriculture

Ethics is by its nature crosscutting and multidimensional, and FAO has therefore designated "Ethics in Food and Agriculture" as a Priority Area for Interdisciplinary Action within its Programme of Work (FAO 2000a), and established an internal Committee on Ethics in Food and Agriculture representing all FAO's technical departments, to guide and coordinate the ethics programme.

The Panel of Eminent Experts on Ethics in Food and Agriculture

The Director-General established a Panel of Eminent Experts on Ethics in Food and Agriculture as of 1 January 2000, for a period of four years, under the terms of Article VI.4 of the FAO Constitution and Rule XXXV of the General Rules of the Organization, and the guidance provided by the FAO Conference, to advise the him on ethical issues in food and agriculture, including forestry and fisheries, in order to better meet country demands for policy advice, and to further global debates on questions related to ethics in food and agriculture (FAO 2000b). The Panel is composed of eight Eminent Experts designated in their personal capacity by the Director-General, in accordance with established procedures and practices.[1] At the time of preparing this note (early 2004) the Panel was being renewed, for a further period of four years and new members appointed.

The Terms of Reference of the Panel of Eminent Experts on Ethics in Food and Agriculture are as follows:

1. *The Panel of Eminent Experts shall reflect and promote reflection upon ethical issues arising out of food production and consumption as well as agriculture development, including forestry and fisheries, in the context of food security and sustainable rural development against an environment of rapid global change;*

[1] The members of the Panel for 2000–2004 were Mr. Francisco J. Ayala, a philosopher and geneticist from the United States of America; Ms. Chen Chumming, a nutritionist from China; Mr. Asbjørn Eide (Chair), a human rights specialist from Norway; Ms. Noëlle Lenoir, a constitutional lawyer from France; Mr. M.N. Salleh, a forest scientist from Malaysia; Mr. Mohammed Rami, a fisheries expert from Morocco; Ms. Lydia Margarita Tablada Romero, a microbiologist from Cuba; and Mr. Melaku Worede, a geneticist from Ethiopia.

2. *The Panel shall in particular consider ethical issues relating to the interests of present and future generations regarding the sustainable use of natural resources, the safeguarding of biodiversity and the balanced mix of traditional and modern technologies to increase food security and sustainable agriculture.*
3. *Based on the above considerations the Panel shall:*
    a. *Promote an overall sense of international responsibility with regard to the development of necessary policies and instruments aimed at maximizing global benefits, while minimizing risks arising from the application of modern technologies to food and agriculture;*
    b. *Seek to increase the awareness of States, intergovernmental organizations, non-governmental organizations, civil society and public opinion worldwide, with regard to ethical issues in food and agriculture in order to promote international understanding and appropriate action on such issues, bearing in mind that different communities have different cultural values;*
    c. *Advise on possible international, regional or national action or the preparation of instruments, as appropriate, to best respond to ethical issues arising out of food and agriculture, with due regard to interdependence among generations, among countries and between food security and other community needs;*
    d. *Encourage exchange of information on all issues of an ethical nature arising out of food and agriculture;*
4. *The Panel may consider any other issue related to the above;*
5. *In its consideration of the above, the Panel shall advise the Director-General on the possible role and policies of FAO.*

The Panel held its first session in Rome from 26–28 September 2000 (FAO 2001a). It discussed general questions identifying ethical issues in food and agriculture, and specifically questions related to biotechnology, including genetically modified organisms. It decided that the ethical foundations of its work should be the Universal Declaration of Human Rights and sustainable use and conservation of natural resources. The most urgent ethical task was to assess activities relating to food and agriculture in the light of their actual and potential impact on the reduction of poverty, hunger and malnutrition.

The Panel held its second session in Rome from 18–20 March 2002 (FAO 2003b), at which it discussed six main themes:

- Ethical aspects of the global emergency of hunger
- Ethical limits and challenges in the intensification of agriculture
- Ethics and economic globalization in food and agriculture
- Sharing the benefits of biotechnologies as part of the advancement of science

- The Trade Related Intellectual Property Agreement (TRIPS), the Union for the Protection of New Varieties of Plants (UPOV) and Farmers' Rights and
- Genetically modified organisms and ethics in decision-making: participation and transparency

The following are matters that the panel has considered:

Ethical Issues in Food and Agriculture

The ethical concerns related to food and agriculture identified by the Panel are essentially twofold: to promote conditions by which sufficient food is produced and where everyone has access to adequate food; and to promote policies and measures ensuring the ecological sustainability of food production, including fisheries, and to ensure similar sustainability in the practice of forestry.

The Panel identified a number of ethical challenges, including loss of biodiversity; the widening gap between rich and poor; and the lack of global governance instruments through which to promote the positive effects of economic globalization and address its negative effects. The Panel recommended the adoption of comprehensive ecosystem management at all levels, whereby natural resources may be managed for multiple objectives – including economic, social, ecological and dietary objectives – taking both long-term and shorter-term interests into account. Buffering the negative consequences of agricultural intensification – including monoculture and overuse of chemicals – was also seen as a very important task, as was counteracting the negative consequences for agricultural research of the concentration of economic power. The Panel also expressed concerns about the concentration of proprietary rights and the overextension of intellectual property rights.

The Panel also recommended greater promotion and protection of all human rights, especially those of the poorest and most disadvantaged, including the right to food, education and information, as well as political rights, and more participatory and empowering approaches to development.

The Panel endorsed the following guidelines and measures, directed at the national and the international level, for implementing its recommendations:

- Creating the mechanisms necessary to balance interests and resolve conflicts
- Supporting and encouraging broad stakeholder participation in policies, programmes and projects
- Designing incentives that will encourage individuals, communities and nations to engage in dialogue and, ultimately, to do what is ethical
- Ensuring the transparency of information and decision-making;
- Fostering the use of integrated and empirical science and technology in the service of a more just and equitable food and agricultural system
- Encouraging cooperation and solidarity among institutions engaged in research and development, making it possible to take appropriate action more quickly
- Ensuring the incorporation of ethical considerations in all programmes, policies, standards and decisions, thereby contributing to improved human health and well-being and environmental protection

- Developing codes of ethical conduct where they do not currently exist
- Periodically reviewing ethical commitments and determining whether or not they are appropriate on the basis of new knowledge and changes in circumstances

Biotechnologies, Including Genetically Modified Organisms

The Panel agreed that science and technology have provided great benefits in the past and are likely to do so in the future, as long as they are properly managed and applied. It noted, in this connection, that international human rights stipulate that everyone has the right to share in the benefits of scientific progress and its applications (United Nations Organization 1948).

The accelerating rate of development within modern biotechnologies, including genetic engineering, has led to intense public debate, especially about its risks and potential benefits. The Panel considered that biotechnologies – in the wide sense – provide many alternatives to the production of genetically modified organisms (GMOs), and that such alternatives should be preferred where there are significant risks or uncertainties at present concerning the use of GMOs, particularly when the expected benefits of the latter are few. Preference should be given to the most appropriate technologies. The Panel considered that ethical challenges were posed not only by technologies themselves, but also by the ways in which they was marketed, utilized and controlled. The Panel was concerned that these technologies might lead to over-concentration of economic power and that poor farmers would be excluded from benefits. The Panel was also gravely concerned with Gene Utilisation Restriction Technologies (GURTs), which it felt might unduly infringe on Farmers' Rights.

The Panel explored ways in which the potential of modern biotechnologies could be realized and the risks avoided. It thought it was misleading to lump together the various biotechnologies under the word "biotechnology". There are, it felt, many different biotechnologies, some modern, others old and tested, some controversial, others not tested at all.

The Panel considered that a key issue was how access to beneficial technologies was facilitated and how benefits from science and technology were shared. It noted that most research was undertaken and most technologies were produced in the North, while the genetic resources were mainly found in the South. Equitable benefit sharing between North and South was therefore crucial, in a system that recognized not only the rights of the researchers but also the rights of all humankind to enjoy the benefits.

The Panel also discussed the impact of the Agreement on Trade-related Aspects of Intellectual Property Rights (TRIPS) under the World Trade Organisation, the system of plant variety protection provided by the International Union for the Protection of New Varieties of Plants (UPOV), and Farmers' Rights. Concerns were raised concerning: (1) the increasing transfer of important knowledge from the common domain (public goods) to the private domain; (2) possible negative impacts of TRIPS on the livelihood of poor farmers; (3) its uncertain impact on sustainable access to affordable, safe, nutritious food for consumers with limited income; and (4) its environmental impact, including its effect on biodiversity.

The Panel pointed to a number of important principles that could guide ethical decision-making regarding GMOs: public participation in decision-making through stakeholder consultations and public debates; access to information, including labelling and traceability standards; and monitoring and oversight processes to ensure public accountability. The Panel considered that a more holistic approach to decision-making, including socio-economic considerations, might result in a more accurate consideration of costs and benefits in the regulatory decision-making process.

Intensification of Agriculture

The Panel also took up the subject of our conference at Wye, the intensification of agriculture. It noted that agricultural intensification – while a necessary condition to produce more food – does not by itself ensure access for all to sufficient and adequate food. It therefore considered that the concept of sustainable development might be broadened to cover ethically sound development, embracing social and ecological sustainability. Intensification should be pursued in ways that are socially responsible and respect the interests of future generations, and pay more attention to the needs of those who are marginal or weak. It considered that the main priority at the present stage was not a global increase in food production, but to ensure broader access, including by those who do not now have adequate food because they cannot afford to buy it and cannot produce it themselves. Intensification that could increase the income of those farmers who are presently marginal and poor would serve that purpose and should be welcome.

Ecologically sustainable intensification would require greater productivity without dependency on non-renewable resources. There was a serious risk that – without more effective management of land and water resources, and of forest and fishery resources – we may be nearing carrying capacities at local, regional and even at world level, sometimes as a result of capital-intensive exploitation that undermines the livelihood of local populations. The Panel also pointed out that, in many cases, smaller scale farmers play the role of farmer-curators, who pursue more traditional farming practices in ways that assist in maintaining biodiversity, local knowledge and technologies, and sustain the traditional cultures of the societies where they live. There may be a need for both, but the farmer-curator may need financial and other support in order to avoid or get out of poverty.

Economic Globalization in Food and Agriculture

In its discussion on globalization, the focus of attention for the Panel was on processes of economic globalization in food and agriculture. The main features were the expansion of foreign private investment in agriculture, food processing and marketing, to a large extent but not only through transnational corporations, and increasing international trade in food, facilitated by the reduction of trade barriers. The consequences of these processes for the environment and for the livelihood of those who suffer from hunger and malnutrition needed to be examined.

The Panel observed that economic power is becoming more concentrated: the world's 200 largest transnational corporations now accounted for a quarter of the world's economic activity, and were often beyond the control of the governments in the territories in which they operated.

One prominent aspect of contemporary processes of globalization was the transition from public services and regulation to privatisation and deregulation. Another was a reduction in the space for national, democratic governance. The Panel considered that the market should not be the sole governor of social and economic processes. It felt that there should be movement towards a global society that gives equal opportunities for all, rather than focussing on economic globalization that gives the prime benefit to those who are already the strongest actors in the global economic system. Global governance was required in order to advance the benefits, particularly for those who are presently disadvantaged, and to offset the risks and negative consequences.

The Panel considered that it might be advisable to develop a code of conduct for economic globalization to avoid some of the negative consequences and to ensure a better and more equitable sharing of benefits for all.

Publications on Ethics in Food and Agriculture

The *FAO Ethics Series* was launched in 2000 and aims at publishing one paper per year. Its main purposes are to promote ethical reflection and constructive dialogue, and clarify the issues underlying current public debates in ways that help both policy makers and the public at large to form their opinions.

Ethical Issues in Food and Agriculture (FAO 2001b) aims to identify the main ethical issues and challenges, which include achieving the global goals of food security for all; achieving sustainable rural development; ensuring equitable participation and balancing the interests of the wealthy and the poor; and guaranteeing viable options for future generations. The paper provided an assessment of current trends, identifies the main issues and stresses the need to balance interests, resolve conflicts and build a more equitable and ethical food and agriculture system.

The stubborn persistence of hunger and poverty raises some of the most burning ethical questions of the age. Freeing humanity from hunger and malnutrition is a moral obligation that weighs more and more heavily as our capabilities and technologies advance. The technological advances and organizational changes affecting food and agriculture systems over the past years have been both radical and rapid, will have repercussions for a long time to come, and have sometimes had irreversible consequences.

Genetically Modified Organisms, Consumers, Food Safety and the Environment

(FAO 2001c) discusses many aspects of the global debate on genetically modified organisms (GMOs), in the light of relevant current knowledge about food safety and environmental conservation. It seeks to unravel and explore the claims and counterclaims being made in the GMO debate from an ethical perspective, considering the proprietary nature of the tools used to produce GMOs, the potential consequences of their use in intensifying food production, and the unintended and undesirable effects that their application could have, both now and in the future.

Both the potential contributions of genetically modified products to world food production and the possible risks with regard to food safety and unpredictable environmental hazards – the most commonly cited being the feared transfer of toxins or allergens and unintended negative effects on non-target species – need to be taken into account, as well as possible long term consequences, such as diminished biodiversity through the loss of traditional crops. Finally, like all the new technologies, GMOs are merely technical instruments that can be used for good and for bad. How they are used will determine whether they benefit the neediest, or privilege specific groups that hold political, economic and technological power. The main beneficiaries to date have been the private sector technology developers and large-scale agricultural producers, mostly in developed countries. To ensure that benefits are shared more fairly with developing countries and resource-poor farmers, intellectual property rights should not impede the use of technologies by those who most need them, and research must be directed towards their needs and the needs of disadvantaged farmers, and ways must be found to guarantee that increased production benefits accrue to the poor and food-insecure.

The Ethics of Sustainable Agricultural Intensification (which was the main document of our Conference at Wye, and which was published in 2004) proposes approaches to analysing the ethical issues surrounding modern agricultural intensification, in particular how sustainability can be ensured. It will focus on informing decision-making including ethical decision-making, using a variety of approaches, including utilitarian, rights-based and virtue-based ethics.

Three further papers are in preparation for publication in the 2004–2005 biennium.

Ethics and Globalisation of Food and Agriculture will seek to explore the term, "globalisation" and to distinguish which major current changes are attributable to "globalisation" and which to other factors. It will discuss ethical issues relating to effects of globalization on hunger, poverty, diversity and equality.

Global Public Goods in Food and Agriculture will address ethical issues relating to the sustainable management of global public goods in the modern world. Among trends discussed will be the progressive privatisation and ownership of natural and intellectual resources. The paper will consider the ethical dimensions of the management of Global Public Goods, as a common responsibility of mankind, and in order to safeguard inter- and intra-generational equity.

Ethical Issues in Fisheries will focus on selected issues in fisheries, including the major ethical issues and challenges posed by multiple use of, and multiple access to, marine resources, including over-fishing, bye-catch, discards, degradation and pollution of the aquatic environment, industrialisation of fisheries and aquaculture, new technologies and the information revolution, and the conflicts between global and local fishers, traditional and modern.

In late 2003, a second publications series was established, entitled *Readings in Ethics in Food and Agriculture*. It will draw together relevant publications resulting from the regular work programme of the Organization. The first publication in the series is the *Report of the FAO Expert Consultation on Food Safety: Science and Ethics*, which was held in Rome, 3–5 September 2002 (FAO 2003a).

Applying Ethical Analysis to Food and Agriculture[2]

It is far more difficult to apply ethical analysis to food and agriculture than in fields such as medicine and health, for a number of reasons. *A priori*, medicine and health lead directly into questions of life and death, sexual behaviour and reproductive health: all raise highly charged philosophical, cultural and religious concerns. Ethical concerns in medicine can also frequently be understood in terms of the individual, and his or her rights, duties and hopes. There are few matters of such immediacy in food and agriculture, nor, in most cases, is the proximity of the actor to the outcome of the action as close: while a doctor must make fraught decisions in difficult circumstances, with immediate results for the patient, few actors in farming system are as close to the results of what they do. Our food and agricultural production and distribution systems are complex, with long chains of actors, often only loosely bound one to another; it is often difficult to describe and articulate the relationships of the actors, and to try to understand their individual responsibilities.

In medicine, moreover, the ethical questions that arise may most often be addressed *within* the medical practice itself, that is, the remedy often lies in the hands of those evaluating the dilemma. This is, of course, not always the case, as medicine is not a walled-around autonomous subject: for example, the question of access by developing countries to medicines critical for tropical diseases and to combat the flail of AIDS is not endogenous to medicine, but depends on decisions regarding intellectual property, market regulation and trade policy, which are outside the immediate control of the medical sector. Clearly, however, actors in the medical sector, supported by public opinion mobilised around objectives formulated in ethical terms, are able to effect decisions in these other sectors.

[2]The reflections in this section are my own and do not necessarily reflect the opinions of the FAO or its members.

In agriculture, the levers of policy and practice are even more frequently held outside the sector, so that the remedy is yet more distant from the practitioner, a trend that will increase with the decline in relative importance of the agricultural sector as economies grow. In industrialized countries, agriculture is a relatively minor sector. For example, in the United Kingdom, agriculture accounts for 1.4% of GDP, and produces about 60% of food needs with 1% of the labour force. (This, however, does not mean that the average farmer is well off, or his or her life easy. A farmer's life is not easy: British farmers are reported to have one of the highest suicide rates of any occupational category.) Increasingly, the decisions that condition agriculture are being taken outside the agricultural sector, for understandable reasons. To continue the British example, the foot-and-mouth crisis of 2001, with the massive slaughter of stock, cost the farming industry £3,000 million in disease control and disposal of stock, some of which was recovered by farmers in compensation, but the rest of the rural economy, particularly the tourist industry, lost an estimated £5,000 million, little of which was ever recovered by those involved. Responsible decision-making in such circumstances cannot purely be the prerogative of agriculturalists.

The long-term perspective is for a relative decline in agriculture everywhere. Although in developing countries, the current relative size and strategic importance of the agricultural sector is much greater than in developed countries, economic development aims at increasing the overall GDP, within which the demand for agricultural products is relatively inelastic, and the part of agriculture ever smaller.

In the more developed countries, this decline is already reflected in the political process. One or two generations ago in Western Europe, many countries had formal farmers' parties. These have almost completely disappeared, and a whole new sector of voter opinion around environmental factors has appeared, reflected in the emergence of "green" parties. There is, in such countries, a still unresolved tension between agriculturalists and environmentalists: on the one hand, agriculture is sometimes seen as in opposition to "nature", but, on the other hand, there is a growing understanding of and support for the role farmers play in maintaining the landscape and providing environmental services. This tension is sometimes formulated in ethical terms.

This does not mean that ethics has little place in food and agriculture. Food and access to food, land and other means of production and ownership, have been among the main nexuses of idea and morality in many cultures, and colour our languages with metaphor even in the most urban societies: food, as Claude Lévi-Strauss said, is good to think with. In most societies, particularly in developing countries, agriculture is one of the key pillars of social organization. Historically, in pre-industrial societies, the ownership and exploitation and exchange regimes in agriculture *were* the structure of society. Marx, for example, derived a whole pre-capitalist structure of society from the control of irrigation waters, "hydraulic" or "oriental" despotism, which Wittfogel turned on its head, to describe the centralising despotism of Marxism. Since the industrial revolution, and with the continuing falling price of transport, the key driving factor in globalisation, economies and technologies, including agricultural technologies, are changing ever more rapidly throughout the world. Times of rapid change are usually times of great opportunity, but the benefits of change often largely accrue to small groups. Understanding these processes, and protecting the interests of those who risk marginalisation with such changes, are ethical challenges.

Food itself is deeply associated with culture, be it through the adaptation of food and culinary culture that the French sum up in the concept of "terroir", or through the food taboos of a number of religions. Cooking and commensal eating are uniquely human habits, which colour our concepts of the gift, of generosity, of kindness and appropriate behaviour: they cannot fail to have ethical undertones.

Important ethical considerations arise from the fact that agriculture is a major consumer, modifier and, in many cases, maintainer of natural resources. Agriculture is the largest single user of water resources, for example. It mines and uses exhaustible mineral fertilisers and is an important consumer of energy, including fossil fuels. There is therefore an ethical imperative to ensure that agriculture uses these resources wisely, to feed us now and to feed the world of the future. Even certain of our renewable resources now approach our have gone over their sustainable maximum production. Many capture fisheries are in this position, requiring difficult national, regional and international policy decisions, and even more difficult policy implementation. Both the impacts of over-exploitation, and the difficult decisions required to rectify the situation, immediately effect both the environment itself, as well as producers (in this case fisher folk and fishing communities, whose livelihoods are at stake) and consumers. Clearly, where there are difficult choices and conflicts of interest, there are ethical considerations.

It is wrong to see agriculture only in terms of problems. Agriculture is a creative and constructive endeavour. Ten thousand years of agriculture since the Neolithic revolution have reformed and recast the natural landscape in the service of humanity, frequently creating increasingly productive farming systems, and maintaining these in a sustainable way. Agriculture, in fact, directly manages about two-thirds of the world's land surface, under an astounding array of ingenious production systems, including, for example, the flooded paddy terraces of Indonesia, the cattle pastoral systems of the Masai and other peoples in Africa, the long-distance *qanat* underground water transportation systems of the Middle East, sheep transhumant systems between the low and high lands surrounding the Mediterranean, and crop and animal rotation systems in many areas of the world. The creative stewardship of agriculture in the environment has in itself strong moral overtones.

We need to recognize that farmers have not merely exploited nature, but have added value to it. An excellent example is found in genetic resources for food and agriculture, that is, in the complexes of domesticated plant and animals, and in their genes. These are not merely a natural resource, like wild plants and animals, because they have been managed, modified and maintained by farmers throughout the world since agriculture began. Unlike wild diversity, it is the diversity *within* crops and domestic animals, built up by generations of farmer selection, not the number of species, that matters.

Beginning some 10,000 years ago, there were a number of independent Neolithic revolutions around the world: for example, barley and wheat were domesticated in the Near East, rice in South-East Asia, the potato in the Andes, millet and sorghum in Africa, and maize in Central America. In each case, local farming communities selected within the genotypes of the original wild plants, farmers substantially altered their nature and created crops. They had, for example, to stop the seed-heads 'shattering' and scattering their seed before harvest and, in so doing, selected for something that is deleterious in the wild plant. It is not surprising that many domesticated plants cannot survive in nature: maize, for example, with its very tight ears, cannot seed itself. If you compare it to wild teocinte, from which it originated, maize is almost unrecognisable, so far have farmers taken it from its origins.

From the earliest days, agriculture has been characterised by the exchange of crops and farm animals, which often spread into new environments. Where possible, farmers adapted them became to new stresses, new climates and new human needs, leading to ever increasing diversity, which swelled the overall portfolio of genetic resources. It is this diversity, maintained as a pooled good, within traditional farming systems, which provides the raw material we need for further agricultural development, and to face up to such challenges as global warming.

From the end of the 15th century, the extraordinarily valuable crops of the Incas and the Aztecs (maize, potato and tomato, and many others) came to Europe, Africa and Asia, while, in the other direction, the crops of Europe, Africa and Asia enriched South America. New centres of diversity grew up outside the centres of origin: South American beans, for example, have a major, secondary centre of diversity around the great lakes of central Africa. As a result, these resources have been very widely shared: wheat, maize, potatoes, yams and rice are now world crops. In our interdependent world, 20 major food crops provide 90% of our food from plants: these are banana, barley, beans, cassava, groundnut, maize, millets, oats, oil palm, peas, potato, rapeseed, rice, rye, sorghum, soybean, sugarcane, sweet potato, wheat and yam.

If contributing to the greater good of the greater number is a moral activity, then farming has a strong ethical component. However, in this time of massive change and specialisation in production systems, there is a new technical challenge to ensure that we do not, through individual inaction, or the inability to find agreed ways of doing so, lose the inheritance of the past, which we need to ensure food for the future. In effect, the modernization of agriculture has led to progressively greater areas being planted to increasingly uniform crops, and to the use of a few highly productive farm animals. There is also an increasing reliance on seeds and animal stock produced by specialist breeders, in both the public and private sectors, and as a consequence the displacement of traditional crop varieties and animal breeds, because seeds are no longer being saved for replanting, and traditional animals are replaced by a small number of breeds. We risk the loss of this diversity, as farming modernizes, unless specific measures are taken to conserve the genes in farming systems, or *ex situ* in crop gene banks and conservation herds. The problem may be technical, but to fail to resolve it, would be an ethical failing, because of the consequences.

How, then, can one usefully apply ethical reflection to food and agriculture, bearing in mind, amongst other things, that few agriculturalists have by their training acquired a formal grounding in philosophy and ethical theory? (This does not, of course, mean that agriculturalists are amoral. Like Molière's Monsieur Jourdan, who realised that "For more than forty years I have been speaking prose without knowing it", the nature of their enterprise has often required them to weigh ethical considerations and take ethical decisions on the fly.) From the beginning of the FAO ethics programme, as shown in the studies in the *FAO Ethics Series*, we have sought to avoid prescriptive approaches and have concentrated on developing ways of approaching specific problems, and above all on developing the tools of ethical discourse in relation to food and agriculture, including through considering how best to relate ethical discourse to other modes of discourse, such as economic and development theory. Our aim throughout has been to encourage decision-makers and the informed public tease out specific ethical questions, so as to help them taking the decisions that are their responsibility.

One major challenge is to identify what precisely the important ethical questions in a particular subject are. If this is not done, then one risks remaining in broad generalities, with little practical outcome. Clearly, not every question benefits equally from ethical analysis: many can be approached at a pragmatic or technical level. In this, it is useful to try to articulate *why* and in *what way* a particular question has ethical implications. One is otherwise likely to label as ethical problems, without real justifying, what are merely received opinions and prejudices. This exercise is particularly important in culturally mixed groups, both to find common ground, and to escape reductive moral relativism and political correctness. Finally, in the complex chains of actors involved in food and agriculture, it is important to identify the actors, their relationships one to another, and their various responsibilities. If ethical reflection suggests that something needs to be done, then who *can* do that something, because only those who can, *should*, which is the ethical imperative.

In the various studies we have prepared, we have sought to lay out the questions addressed so that the reader can follow the chain of reasoning, and make autonomous decisions as to its validity. We have also tried to recognise and show the arguments on the question under discussion that have been put forward by a wide range of players: for example, in the study entitled *Genetically Modified Organisms, Consumers, Food Safety and the Environment* we deliberately set out to provide tools to understand and evaluate the very discrepant statements that different persons and constituencies had made on these subjects.

Similarly, we had a number of objectives in mind in addressing *The Ethics of Sustainable Agricultural Intensification*. The first was methodological, to demonstrate, without attempting to be exhaustive, *how* ethical analysis could be applied to a complex problem, and to propose a toolbox, however simplified, for doing so. The second was to discuss how the ethical problems inherent in the concept of "intensification" might be addressed: in this approach, we were aware that the very concept of intensification is tendentious, with some constituencies arguing against intensification as being inimical to social justice for small producers, and others arguing for intensification as a good in itself that overrides considerations of its immediate effects. Briefly, when an idea is tendentious, it is susceptible of ethical discourse. Finally, we deliberately wished to introduce into the argument a historical element, for a number of reasons. One was to combat the tendency of technical specialists to address current problems as if similar situations had not occurred in the past, from which one might learn. (Economists and development theorists, perhaps because they look forward in their work to changes they hope to effect, are particularly prone to this.) If the tendency is not combated, there is the risk of for ever rediscovering what a modicum of historical understanding will supply: this is what Sorokin (whose target was sociological theoreticians) called the myth of the "New Columbuses", repeatatively rediscovering the Americas.

There are other important benefits from analyzing a question in terms of what Braudel called *la longe durée*. It encourages balance and humility, as well as an appreciation for real achievements. It is important to realize how very recently even developed societies, such as those of Western Europe, have escaped the cycle of famine (war-caused famine not included): this was only in the third quarter of the 19th century in many places. Braudel's careful description of cyclic famine and plenty in the Mediterranean world in the 14th to 16th centuries is salutary reading. A focus on time was also intended to point up how unusual the present moment is against the backdrop of millennia: we are in an unusual moment of both rapid population growth and of rapid production increases, (seen globally) that so far keep up with population growth. As the 1996 World Food Summit *Plan of Action* recognized (FAO 1996), the

> *Availability of enough food for all can be attained. The 5.8 billion people in the world today have, on average, 15 percent more food per person than the global population of 4 billion people had 20 years ago. Yet, further large increases in world food production, through the sustainable management of natural resources, are required to feed a growing population, and achieve improved diets. Increased production, including traditional crops and their products, in efficient combination with food imports, reserves, and international trade can strengthen food security and address regional disparities.*

Moreover, a historical approach in itself raises ethical problems. One is judging the actions of the dead, that is, judging actions where there is no possibility of modifying the outcome. Another is evaluating the morality of actions with the hindsight of their outcomes, intended and unintended. As the results of actions in the past are absorbed into longer-term historical processes, how far can we judge of their morality? Take for example Alexander the Great's conquest of the East. The human suffering that resulted was enormous. To take one small instance, Alexander's army three times abandoned to their death its camp followers with the soldiers' children by them. And yet, when we look at the episode in terms of its effect on history and the creative cross-fertilization of western and eastern cultures, can we "absorb" such momentary evil in the long-term good? Does the question make sense, and if so why? These are questions we wanted to raise in the context of agricultural intensification, to deepen the discussion. They are not merely arbitrary, or valueless because an unchangeable aspect of the past. They apply directly to such arguments as that harm must be done to some group, or to the environment, because the long-term benefits justify this. Many such arguments have recently been made in neo-liberal economics, in relation to matters such as structural adjustment, trade liberalization and globalization. How can we evaluate them, including in ethical terms? Economists will do so against projected, modeled outcomes. Historians will bring the corrective of what actually happened in the past. Ethics must encompass both, so that wise decisions can now be taken.

Chapter 3
History, Ethics, and Intensification in Agriculture

John H. Perkins and Rachael Jamison*

Every agricultural scientist knows that his/her specialty underwent stupendous changes between 1900 and 2000. What a century it was for the development of knowledge and a prodigious increase in the output of food and fiber for the world's growing population.

What is murkier is how ethical problems underlie the startling changes in agricultural practices and outputs. How, a good scientist might ask, can any ethical problems be associated with knowledge that can end famine and hunger? Surely it would have been immoral not to develop these sciences.

Yet, often to the consternation of agricultural scientists, critics of high yielding agriculture have launched numerous ethical and moral attacks against the new practices. Are these charges true? And if so, what can agricultural scientists do about them?

This chapter summarizes important historical trends affecting particularly interesting cases of agricultural change. The main argument is that science and technology are necessary but not sufficient to understand agricultural change. Other political economic, cultural, and ethical factors are also important. Scientists need to understand these issues surrounding agricultural change as they plan their future research.

Agriculture and the Yield Transformations

Many remarkable changes in history transformed human links with their food supplies. In the Neolithic, beginning about 10,000 years ago, humans firmly embraced agriculture. In so doing, people acquired new ways of procuring their food, ways that ultimately led to new cultural patterns and civilization itself.

The Neolithic revolution was merely the baseline for later and even more profound changes. Yields rose over the millennia to levels significantly above those

* John Perkins is a retired faculty member of Environmental Studies at The Evergreen State College. Rachael Jamison is Coordinator of the Green Building Program for the Washington State Department of Ecology

P.B. Thompson (ed.) *The Ethics of Intensification*,

achieved by our Neolithic ancestors. Until approximately 1,000 years ago, the rise in yields was slow, primarily the product of farmers who came to know how to coax higher yields from their land, seeds, and labors.

Over the last 1,000 years, particularly in the last century, yields exploded. Wheat fields, for example, might produce 0.5–1.0 t per hectare in medieval times, but now some farmers are able to harvest more than 7.0 t per hectare. Similar stories exist for rice, maize, and other staple crops.

Over the centuries, yields rose incrementally with each new practice. A percent or two here, five percent there, and, occasionally, a big ten percent jump was the standard fare. Periodically, a much larger increase in yields resulted from a particular innovation. Moving from a two-course rotation to a three-course rotation in medieval Europe, for example, was an enormous leap of one-third in potential yields from the existing land base.

In contrast to earlier, farmer-invented increases in yield, the events of the last century resulted from the combination of farmers and professional agricultural scientists. Soil scientists, plant breeders, and pest control specialists developed new production methods for nearly every part of the globe. The new agricultural practices currently enable the world's farmers to produce about 2,000 million tons of basic grains annually. Most dramatic were the doubling and tripling of yields that came with the high yielding varieties of wheat and rice during the green revolution of the 1950s through 1970s.

Higher yields brought benefits, but ethical quandaries also appeared. Professional scientists had become co-responsible with others for these ethical consequences.

Intensification and Ethics

Agriculturists, and their allies in science and technology, were ecstatic about the ability to get more yields from land. Advocates for the new agriculture hailed this intensification of land use as the vehicle for prosperity, security, and health for people and their livestock.

Although intensification was welcomed, it was also irreversible and unequal in its effects. Changes that cannot be reversed and that create uneven effects have ethical consequences. *Intensification*, therefore, has not been an unqualified benefit. Inventors (originators) and innovators (adopters) of agricultural change have seldom given much attention to the complex consequences of change and their ethical implications. For this reason, it's important to delve more deeply into what the concept means and how it has been a part of agricultural history.

Intensification has been defined by the "FAO Ethics Paper" (Chapter 1) to be

> … an increase in agricultural production per unit of inputs (which can be labor, land, time, fertilizer, seed, feed, or cash). For practical purposes, intensification occurs when there is an increase in the total volume of agricultural production that results from the higher productivity of inputs; or maintenance of agricultural production while decreasing some inputs (such as by more effective delivery of smaller amounts of fertilizer, better targeting of plant or animal protection, mixed or relay cropping on smaller fields).

Three *caveats* about this definition of intensification must be noted, although they are only minor constraints on its utility. First, as others have noted, it's extremely rare to see an intensification of one input without a shift in the uses of others. If, for example, labor power is amplified through machinery, then uses of some other inputs are also likely to change. The definition, however, speaks only to the more efficient use of one input at a time.

Second, the closeness of *intensity* to *efficiency* may be useful and important, but it can also be distracting or misleading. Adulation of efficiency in market-based societies means that normative virtue may be unconsciously associated with any practice that makes the use of a scarce or valuable input more intense or efficient. Vigilance must be maintained to make sure all ethical issues are examined, not just assumed.

Finally, the draft definition has a partial list of inputs, but it is silent about the kinds of outputs valued from "agricultural production." Production may be measured in physical, economical, or chemical/biological units. The historical episodes of intensification sketched below involve all of these units. The choice of particular units to measure production may itself involve significant ethical issues.

Intensification of land use was the most or one of the most important processes of change. Yields measure the effectiveness of land intensification, which is usually expressed in kilograms of useful crop output per hectare of land input. Fertilization, irrigation, pest control, and the use of improved varieties produced most of the intensification of land.

Intensification of labor power became at least co-equal to intensification of land during the 19th and 20th centuries, especially in the industrialized countries. This aspect of intensification is usually measured by outputs of crop product per input of human labor. Mechanization provided the major avenue for intensification of human labor power.

As the examples of intensification of land and human labor suggest, the higher outputs now obtained for these two inputs are the result of new technologies. Modern science, from the 17th century on, provided a helpful adjunct to invention of new technology. Chemistry, physics, math and statistics, biology, the engineering sciences, and the agricultural sciences each allowed a more systematic understanding of the material world. Science and technology weren't, and aren't, the same, but they reinforced each other.

Increased yields from land, with less human labor, were the objectives of formal agricultural science research, and these objectives were reached. The natural environment became more favorable for humans. Food and fiber yields were more abundant, stable, and of higher quality. Population grew as people were better supplied with essentials.

Unfortunately, other consequences also attended the intensification of land and human labor. These secondary effects were often not predicted, because no one thought about what was being done to the human and natural resources outside of the framework of the primary objectives. Some effects were highly negative, others more benign.

Examples include pollution from fertilizers and pesticides, both of which were important in obtaining higher yields from each hectare and each hour of human labor input. Other examples include the displacement of labor from agriculture by machines that enabled one person to do the work formerly done by 40 or 50.

Technologies that successfully intensified the use of land and labor had a complex mixture of benefits, risks, and costs. Studies of particular technologies and their impacts have blossomed, aided by professional scientists. Unfortunately, agricultural scientists have generally not been much involved with the assessment studies. Thus many remain challenged to understand the complexities of the fields they have chosen for their careers. The discussion below seeks to help agricultural scientists see these impacts in a larger context.

Intensification Before 1960

People living in industrialized countries after 1900 drew their sustenance almost entirely from agriculture. Only a bit of fish, wild game, or the gathering of wild berries and nuts remained of what was once the only way of life: hunting and gathering. Yet for most of human evolutionary history, our species was a hunting-gathering creature. How, where, why and when H*omo sapiens* began to shift to agriculture stimulated an extensive literature and a vigorous scholarly debate.

Many creation stories describe the transition to agriculture. For example the Bible (Genesis, chapters 2 and 3) offers one of the earliest explanations: Eve's deception by the serpent in the Garden of Eden brought forth the wrath of God against Adam and herself. As they were tossed out of paradise, God told Adam that he was henceforth doomed to scratch his living from the ground, which would bring forth thorns and thistles as well as the herb of the field. "In the sweat of thy face shalt thou eat bread…" In this rendition, the start of agriculture was not a joyous occasion (Genesis 3.19; Tudge 1998).

Scientific studies during the 20th century generally followed one of three paths to provide an explanation of agriculture's origins. An early school argued that the invention was largely a step of progress that occurred first in southwest Asia or northeast Africa with the domestication of barley and wheat (see, e.g., Peake 1928). Later studies, based on more extensive work in paleobotany and archeology, suggested multiple centers of origin in South America, Mesoamerica, southwest Asia, and East Asia.[1]

Most intriguing, however, is Colin Tudge's speculative essay that in fact *Homo sapiens* became "proto-farmers" in the late Paleolithic, about 40,000 years ago. Proto-farming people manipulated the environment (plants and animals) in a way to encourage the species they found most beneficial. Tudge argues that what is often called the "Neolithic Revolution" was actually the end point, not the beginning, in the transformation of people from hunter-gatherers to farmers, horticulturalists, and pastoralists (Tudge 1998). Bogaard enriches this argument by maintaining

[1] See, for example, MacNeish (1992); and Cowan and Watson (1992).

that archeological evidence points to intensive "garden agriculture" as a major step in the Neolithic transformation. Cereal cropping integrated with animal husbandry formed the agricultural revolution in both the Near East and Europe (Bogaard 2005).

Debate on the origins of agriculture will undoubtedly continue, but what is in less doubt is that the Genesis story may have captured the essence of the transition: it more than likely occurred out of desperation as opposed to fulfilling a vision of progress. For millennia after the origins of agriculture and cities, the health and nutrition of most people was not better than hunter-gatherers (Cohen 1989).

Population growth probably drove people out of necessity to invent agricultural technology and become farmers,[2] but once begun relatively stable food supplies allowed further increases in population. With higher numbers of people, no alternative but agriculture existed to produce enough. Only in the 20th century did agriculture achieve yields high enough to markedly improve health (Tudge 1998).

The Neolithic marks an approximate dividing line between people predominately living by hunting and gathering before and agriculture afterward. Yields of agriculture undoubtedly began to rise slowly as people gained familiarity with soils, climate, and plant and animal varieties. Aside from some Greek, Latin, Chinese, and Arabic manuscripts, however, we know little of the technological improvements between about 10,000 years before the present and 1,000–2,000 years before the present.

Remnants of Byzantine, Islamic, and Western European civilization in medieval times (1,500–700 years before present) nevertheless indicate that substantial technical knowledge had developed in the 8,000 intervening years. Examples include ways of making wines and olive oil, irrigating dry lands with *qanats* (underground canals) and *saqiyas* (animal-powered waterwheels for lifting water to fields), development of heavy wheeled-plows, adoption of the rigid horse collar allowing horses to replace oxen in plowing, and a shift from two-field to three-field rotations (Long 2003).

Informal, probably artisan-based, invention continued up to the 17th century. In Europe, for example, rotations such as the three-field pattern became more widespread, and more complex rotations such as beans-flax, hemp-grain, and millet-grain emerged in southern Europe. Perhaps most spectacular in opening land and raising yields were the astounding developments in hydraulic technology, such as the reclaiming of lowlands in The Netherlands and the draining of the Fens in England. Better irrigation became common in England and Italy. Social changes such as enclosure in England were the harbingers of industrialization and the emergence of modern capitalism (Long 2000). In short, agriculture by 1600 had moved a long way from its early forms in Paleolithic and Neolithic times.

In the years after 1600, the pace of agricultural intensification took a quantum leap upward. In the subsequent 400 years, the changes in practices far surpassed anything that had been done between 10,000 and 400 years before the present.

[2] This is the thesis of Esther Boserup (1966).

At the core of the new wave of change was the shift from informal to formal methods of invention and innovation. Before 1600 most change was probably introduced by a practicing artisan who knew his or her material intimately, and the names of clever inventors are forever lost in the dust of history. After, a series of transformations eventually placed the responsibility for cleverness in the hands of professional scientists and engineers.

Three intertwined sets of changes in social institutions were involved. First, modern science, fully ensconced in materialism, empiricism, and quantification, replaced earlier inquiries based on assumptions of vitalism, God, and metaphysics. Carolyn Merchant makes an eloquent argument that this change in thought patterns was the "death of nature," or the loss of an image of the earth as an organic, living, female entity. Modern science dealt with dead matter in motion and the search for laws that explained phenomena without recourse to the supernatural (Merchant 1980).

Closely linked to this scientific way of looking at nature came the emergence of liberal democratic capitalism. Individual rights and the rule of law enabled the development of extensive trade and the freedom of people to leave an agrarian, feudal economy for life in the cities. With nature dead, no limits accompanied the quest for increased trade and wealth. Science slowly became an accomplice to the drive for wealth, because science promised, and delivered, greater control over the natural world.[3]

Industrialization was the third foundation of the modern, inventive world. Exploitation of fuels and metals enabled inventors and entrepreneurs to create a world in which energy, machines and chemicals did the bidding of humans.[4] Invention of new methods became the paramount social virtue, and the capitalist state greatly encouraged a torrent of new developments protected by patents. Science oscillated with technology in providing the fodder for urban-based industrial economies, all fed by and dependent upon a highly productive agriculture.

And what became of the old, traditional agrarian life in which 90 percent of the people worked the land and the economy was agricultural? Over a period of 400 years, the labor force working the fields dropped, in some cases, to three percent or less of the population. Agricultural yields skyrocketed during the transformation, but the wealth of the earth was created with vastly fewer hands. A person would no longer live by the sweat of his or her face laboring to harvest wheat for bread.

Between 1800 and 1960, a mere 160 years out of 10,000 years of agriculture, a new method of food production emerged. At its core were institutions fully dedicated to producing new science and technology to increase and stabilize yields. One of the first was the Royal Agricultural Society of England (1840), enshrined through its motto, "Practice with Science" (Russell 1966; Goddard 1981). The

[3]See, for example, discussions on the origins of liberalism and the institutions of market economies and private property: Holmes (1995); Salvadori (1977); and Oldfield (2000).

[4]See, for example, the following discussions on the origins, drivers, and consequences of the emergence of industrialization in early modern Britain and elsewhere: Polanyi (1944); and Beaudreau (1999).

Society sought to increase the prosperity of English agriculture, all with the blessings of the Queen, in the increasingly liberal, democratic society that was emerging from the remnants of British feudalism.

The United States went several steps further in the period 1862–1914. The Morrill Acts of 1862 and 1890 created universities for farmers' children and for educating future farmers. In the Hatch Act of 1888 the USA created research institutions in every state. The final piece, the Cooperative Extension Service, came in 1914 and promised to educate farmers in every county of the land about modern, high yielding agriculture (Cochrane 1993). After 100 years, few farmers were left on the land, and many farm children had headed off to the cities.

The spread of systematic invention of new agricultural practices was made global in the years after the end of the Second World War. Based on models developed by the Rockefeller Foundation in Mexico, the US government launched a major effort to transfer and build institutions for scientific agriculture across the globe. For the US, the moving factor was the desire to defeat communism, but the transfer was embraced by the leaders of every country it touched. After 1960, many national agricultural research and education networks existed, and the World Bank supported an international network through the Consultative Group on International Agricultural Research (Perkins 1997).

Each invention and innovation, however, has its own moral complexities. As agricultural scientists work in the post-2000 period, they need to develop a capacity to understand the kinds of choices that can be made and the consequences that attend every choice. We explore these topics in a series of case studies in the next section.

Selected Case Studies of Significant Intensification Events

Chemicals Used as Pesticides, 1850–1950[5]

Of all the horrors described in the Old Testament, few are as fearsome as Jehovah's order to Moses to bring a plague of locusts upon Egypt:

> And the locusts went up over all the land of Egypt…; very grievous were they… For they covered the face of the whole earth… and they did eat every herb of the land and all the fruit of the trees… and there remained not any green thing… through all the land of Egypt.
>
> (Exodus 10. 14–15)

The story in Exodus is supernatural, but plagues of pests are quite real. A non-supernatural way of dealing with locusts and other pests was one of the most spectacular triumphs of agricultural science and technology. Before 1850, uses had been made of botanical extracts such as rotenone, pyrethrum, and nicotine.

[5]This case study is adapted from material in Perkins (1982).

These products, however, were expensive and tended not to last once exposed to the elements.

Through chemistry and knowledge of pest habits, farmers learned how to put thin films of specific, synthetic chemicals over what was to be protected or killed. Paris green (copper acetoarsenite) was the first of the deadly, new chemicals that was successful. In the late 1860s an unknown farmer discovered this material, invented to be a brilliant green dye, killed Colorado potato beetles in the American Midwest.

The pace of subsequent invention was slow, however, and the next two major breakthroughs were the use of Bordeaux mixture (calcium hydroxide and copper (II) sulfate) to control downy mildew on grapes in France (1882) and lead arsenate to attack gypsy moths (1892) on hardwoods in the northeast of the United States. Lead arsenate quickly found wide use on a number of fruit and vegetable crops, but it was not cheap or effective enough to warrant applications to the major field crops. Calcium arsenate, developed in 1917, was the first synthetic chemical extensively used on a field crop, cotton.

Paris green, Bordeaux mixture, lead arsenate, and calcium arsenate were all based on inorganic materials and the toxicity of metals. As long as the chemical industry was confined to poisonous metals, it was limited in the kinds of pesticidal materials it could produce.

Organic chemists later opened a new world of potential pesticides, and synthetic organic chemicals. By the 1950s, these new compounds replaced the earlier ones. World War I created vast incentives for invention in chemistry, and one of the first successes was *para*-dichlorobenzene (PDB), a by-product of picric acid production for explosives. PDB controlled the peach tree borer with an effectiveness that had never before been seen. Chloropicrin, a tear gas, and a few other minor chemicals were also successful on a smaller scale (Russell 2001; Perkins 1982).

Historian Edmund Russell argues that war and the invention of pesticides were deeply intertwined (2001). This connection was particularly powerful in the circumstances surrounding the invention, development, and adoption of DDT (*p,p*'-dichloro-diphenyl-trichloroethane), the invention that set the pattern for pest control and pesticides after 1939. A German chemistry student, Othmar Zeidler, first synthesized DDT in 1874 as an exercise in chemical synthesis. Although Zeidler had no commercial ambitions for the material, a Swiss chemist, Dr. Paul Herman Müller, later found one.

Müller, working for the Geigy Company in Basel, launched a search for woolen protectors in the early 1930s. In September, 1939, just as the German *Wehrmacht* marched into Poland, Müller announced that DDT was a very potent insecticide that was cheap, easy, safe to use, and very durable when exposed to the elements (Perkins 1982). Safe to use in his context meant that the person exposed to DDT was unlikely to suffer acute illness. Later the criteria for "safety" underwent substantial change.

The Swiss, due to the outbreak of war in Europe, quickly adopted DDT for both agricultural and public health purposes. Swiss officials shared the information about DDT with Germany and the United Kingdom, as well as with the United States. American entomologists, within a few years, had developed the compound as an effective agent to protect soldiers and civilians against malaria-bearing mosquitoes, typhus-bearing lice, and other insect pests.

Once hostilities ended in 1945, American agricultural entomologists quickly developed DDT's uses in agriculture, forestry, and home and garden areas. More importantly, the American chemical industry saw in DDT a vision of a new world of pesticide products. If the DDT molecule was an effective insecticide, how many other molecules could be found that could control insects, weeds, fungi, and many other categories of pests?

For 15 years after 1945, DDT and a host of other new products from the chemical industry remade the human relationship with pests. Killing by chemicals became the mantra adopted by pest control scientists, pest control decision makers, the chemical industry, and public officials regulating these toxic substances. DDT itself became the most widely used product with 188,000,000 pounds produced in the United States in 1962–1963 (Economic Research Service 1969).

Agricultural yields went up under the protective umbrella of pesticides. Produce formerly lost was saved and was of better quality. No longer would the old joke be relevant: "What's worse than biting into an apple and finding a worm? Finding half a worm." (Quality here meant without insect and fungal damage, not freedom from pesticide residues.)

New farming systems became possible. Corn could be grown year after year in the American Midwest instead of in rotation with other crops to control corn root worms. Cotton could be planted closer together, fertilized and watered heavily, and still produce a bountiful crop. Such practices were impossible due to boll weevil, unless pesticides were used. Immense amounts of labor could be saved by killing weeds with herbicides instead of hoeing or plowing. In short, both the productivity of land and human labor expanded rapidly due to the availability of pesticides.

Had these been the sole consequences of pesticide use, no one would have ever complained about them. The trouble came when evidence demonstrated that many other consequences attended the widespread use of synthetic, organic pesticides. Two factors within farmers' fields appeared first. Both were counterintuitive in that the evidence indicated use of pesticides *decreased* yields, rather than the expected *increase*.

First was the appearance of resistance in target insect populations. Although known since 1908 in apple production (Melander 1914), the blow against the new pesticides came in 1954 with the first reports that boll weevil in cotton was no longer controlled by a number of chlorinated hydrocarbon insecticides (Roussel and Clower 1955). Cotton entomologists and farmers alike were in a panic to resolve the matter. Switching to the organophosphate insecticides was the major route to temporary salvation, but that practice, still based on reliance on insecticides, was short-lived. Severe resistance appeared in many other insects on many crops in all parts of the world. Resistance alone was a crisis for chemically-based insect control.

The second alarm came in the mid-1950s (Ripper 1956). Synthetic, organic insecticides were powerful against target insects (at least until resistance appeared), but they were also deadly against the multitude of parasitic and predatory insects that were quite valuable, yet mostly unrecognized, in keeping pest insects in check. Killing natural enemies unleashed the reproductive power of pest species with the result that using insecticides created larger populations of pests. This was called "resurgence."

Even worse was at hand, however. Sometimes the death of natural enemies unleashed a previously innocuous species that suddenly jumped to pest status. Cotton boll worm and tobacco budworm, for example, both pests on cotton, were seldom a nuisance before the new insecticides aimed at boll weevil became standard practice in cotton production. With the death of natural enemies, both boll worm and budworm rose to prominence as "secondary pests." The crowning blow came when an insecticide-induced secondary pest became resistant to the very chemicals that induced it. When boll worms and budworms were resistant, the cotton farmer was in trouble (Perkins 1982).

Both resistance and destruction of natural enemies were so prominent and so obvious, that they quickly captured the full attention of both professional scientists and farmers. Had no other consequences of pesticide use appeared, these alone would have forced changes in pest control practices. But there was more.

In 1962, Rachel Carson published her famous book, *Silent Spring*. Carson, a trained biologist and prize-winning writer, outlined an "ecological theory of pesticide behavior." She combined the phenomena of resistance and destruction of natural enemies with other properties of these chemicals. They were mobile and moved from place of application to many other parts of the world. They were often persistent and built up in the tissues of other, non-target organisms, including people. And they were toxic in ways not anticipated or desired by scientists and farmers who just wanted to kill offending pests.

When President John F. Kennedy noted that he had read the pre-publication version of *Silent Spring* in *The New Yorker*, pesticides became an issue of public policy. Politicians might have been content to let scientists and farmers grapple by themselves with resistance and natural enemy destruction, but they could not sit still and let their constituents and wildlife be slowly poisoned. Fears of cancer induced by pesticides did not make a good case for re-election. That the bald eagle, America's symbol for freedom, was one of several raptorial birds severely affected by DDT, was simply not acceptable. The crisis of pesticides was real, and legislation enshrined environmental protection as the driver of pesticide regulations.

Since 1962, pesticides have continued to be a lightning rod for political controversy around the world. Virtually all countries now have more stringent laws regulating these chemicals. Many of the early chemicals, such as DDT, are outlawed in most countries. Yet pesticides continue to have an allure: they work, they are cheap, and they are more or less safe for workers to apply. Agricultural scientists today, however, have moved to see the problems of pest control in a vastly different light. Future research in pest control must recognize that destruction of pests changes ecosystems, changes human ecology, and leads to unforeseen dilemmas with ethical consequences.

Agroecology and Integrated Pest Management, 1880–1980[6]

Historian Donald Worster traces the roots of modern ecology to the writings of Carl von Linné (Linnaeus), the Swedish botanist and taxonomist, and Gilbert White, an English parson who wrote poetically about the natural history of his village. Neither used the term "ecology" (a word not yet coined), but both understood that nature had order and pattern. Furthermore, both knew that human life was in some way vitally linked to nature's economy (Worster 1994).

Linnaeus sought to explain the order of nature with a taxonomy of all God's creation. Creation was immutable, and for Linnaeus the taxonomic scheme showed the perfection of heavenly works. People, as God's chosen agents on earth, could name all the other species with Linnaeus' scheme.

White, equally devout of God, was more interested in the patterns of relationships that existed between all the animals and plants in his neighborhood. That each depended upon the other and that Man depended on all plants and animals was simply a sign of the harmony and beneficence God created in the world. For White, people needed to understand how species worked together in order to find happiness, tranquility, and virtue.

For about 100 years after Linnaeus and White, little practical use was made of either man's works. Charles Darwin, the English natural historian, complicated the philosophical landscape about nature with ideas that challenged both Linnaeus and White. For Darwin, nature and species were not immutable. Indeed far from it, as he saw natural selection working slowly but steadily to transform one species into another, to create multiple new species out of one, and to drive others extinct. Contrary to White's harmony, Darwin perceived nature as a relentless struggle for survival.

By the 1880s, numerous biologists had developed a literature exploring how nature worked as a whole. Many arguments went on among the biologists, but seldom did they brush up against the work-a-day operations of agricultural production. Farmers were simply not involved with these academic debates.

As scientific research in agriculture became professionalized in the last half of the 1800s, opportunities emerged for biological philosophy to engage the problems of agriculture. As the center of North America was being settled, outbreaks of Rocky Mountain locusts attracted the attention of biologists seeking to explain and predict these wild gyrations in the population size of this pest. The locusts, however, soon disappeared, probably due to destruction of their montane river valley habitat (Lockwood 2004).

Work on the Rocky Mountain locust, plus other pest species in agriculture, concentrated first on basic biological dimensions: how do you identify the species, what is its life cycle, what is its range, what are its habitat requirements, and how does it reproduce? With these questions answered, biologists reasoned, perhaps they could predict outbreaks and control them.

[6]Ibid., pp 61–95.

Agricultural biologists began to link basic biological knowledge with the emerging theory of ecology starting in the late 1800s. Ernst Haeckel, a German biologist, coined the term ecology in 1866 to further develop a concept previously explored by the phrase "nature's economy" (Worster 1994). At the heart of the matter was the concept that promoting the productivity of one species (the "crop" or "livestock" species) depended upon knowing how the desirable species interacted with other species and with its physical environment.

The biological control of insects was born as a scientific specialty by linking basic biology of a pest species with its biological and physical surroundings. Charles Valentine Riley, a US Department of Agriculture entomologist, demonstrated the enormous power of this linkage when he sent Albert Koebele to Australia to search for natural enemies of cottony cushion scale. This insect had entered the newly established citrus orchards of California. When its population blossomed it threatened the citrus growers' viability. Riley was almost certain it had come from Australia, where it was not a problem. Therefore, he reasoned, it was under the control of natural enemies, probably other insects.

Koebele's quest was successful and in 1888 he introduced the vedelia beetle in the citrus orchards around Riverside. Quickly, the cottony cushion scale insect was reduced to non-threatening numbers. From this venture was born a century-long embrace of the State of California with biological control as a way to control pest insects. The US Department of Agriculture shared this embrace through the 1940s, when it lost its prominence to the siren call of the synthetic organic insecticides. In California, however, biological control remained a prominent part of the entomology departments of the University of California at Berkeley and Riverside (Perkins 1982).

Other universities in the United States, and other countries as well, maintained interests in biological control, but the California biological control specialists were exceptional, especially after the new insecticides began to eclipse biological control after 1945. The legacy of this scientific tradition enabled the California scientists to play a leadership role in devising a scheme to extract insect control from the twin traps: resistance to insecticides and the destruction of natural enemies.

Work on alfalfa butterfly from the late 1930s to the early 1950s culminated in the concept of "integrated control." Alfalfa butterfly is a native species that feeds on alfalfa. The moth, however, has considerable variation in its numbers from year to year because of a parasite that sometimes keeps its populations low. Among the first discoveries made about this moth-parasite interaction was that indiscriminate treatment with insecticide could decimate the parasite and leave the moth free to multiply to very large and damaging numbers.

Thus did Ray Smith and Ken Hagen learn that the best control of alfalfa butterfly came from going into the fields and counting the number of caterpillars and its parasites. They then developed guidelines to tell them when to treat with a chemical (because the parasite was not sufficiently abundant) and when to allow nature to keep populations down. This was *integrated control*, and growers were willing to support it.

In 1954, a new insect, the spotted alfalfa aphid, entered California and raised havoc with alfalfa production. The immediate response was insecticide treatment, but by 1956 resistance had set in. Smith and Hagen, along with Robert van den Bosch and Vernon Stern, solved the problem by 1959 with a more elaborate *integrated control* scheme. Their solution included importation of parasites, learning to count pest and parasite in the fields so that they could predict when insecticide might be necessary, and using the insecticide demeton at very low doses so as not to kill parasites. Integrated control, later to become integrated pest management or IPM, was a uniting of ecology, classical biological control, and pesticides.

Various successes followed the work with spotted alfalfa aphid. In the political upheavals following Rachel Carson's *Silent Spring*, support grew in Congress for funding of alternatives to the dominant chemical control protocols. In 1970, a multimillion dollar, multi-university study began to develop IPM for cotton, soybeans, alfalfa, citrus, stone and pome fruits, and pine trees. It was a case where the real problems for the entomologists and their farmer clients centered on resistance and natural enemies. But for Congress it was the environmental pollution and the prospective banning of DDT that pushed the appropriation through.

More generally, textbooks in IPM entomology and agroecology followed. A new generation of students was to see that agricultural production was part of nature's economy. While scientific factors were important in these developments, it is hard to imagine that the science would have moved as far as it did without the surrounding social unrest about pesticides and, more generally, the enormous impacts caused by high yielding agriculture. Agricultural scientists engaged in promoting the development of alternative methods might well consider that support of their endeavors may depend upon the potential to solve larger social problems and not just serve agriculture.

Mendelism, Plant Breeding, and National Security, 1900–1980[7]

For modern biologists, Charles Darwin is as close to a patron saint as is possible to get. Part of the reason for his august status lies in the practical values of Darwin's way of viewing nature. Ironically, evolution by natural selection, the hallmark of Darwin's work, is not often admired for its applied utility. Plant breeders, however, gleaned inspiration from the fact that Darwin's evolutionary theory made *variation* the center piece of biological study.

Before Darwin, variation among individuals of a species was a distraction and a nuisance for biologists. Early plant improvers, such as England's Thomas Knight, John LeCouteur, and Patrick Shirreff for example, had some successes in selecting and breeding new wheat varieties. They had essentially no ability, however, to predict and control the outcomes of crosses or to understand the limits of selecting

[7]This case study is adapted from material in Perkins (1997).

seeds to start new varieties. In short, they did not understand the patterns and order underlying the variations among different wheat plants.

Darwin himself was perplexed on exactly how variation among individuals was passed from generation to generation, but he knew variation was the subject to be studied. Gregor Mendel, a now famous monk of the Austro-Hungarian Empire, proposed a mechanism of inheritance based on factors passed from parents to offspring in regular, predictable ways. Although Mendel's work was not understood for its practical implications in his life time, in 1900 it started to be seen as useful for plant breeding as well as studies in evolution.

Liberty Hyde Bailey of Cornell University coined the term *plant-breeding* in the 1890s and wrote the first text book on the subject in 1891. By 1904, in the third edition of *Plant-Breeding*, Bailey merged Mendel's theory of inheritance with Darwin's concept that variation was the main topic worthy of study. In the ensuing decades with the appearance of new treatises on plant breeding, Darwin's philosophy on variation became implicit, and Mendel's particulate theory of inheritance provided the explicit basis for practical methods.

Even before Bailey codified his thoughts in the 1904 edition of *Plant-Breeding*, English wheat breeder Rowland Biffen of Cambridge University began to put Mendel's theory to use. Biffen saw that Mendel's scheme offered guidance on how to select parents for crosses in the quest to find novel, useful varieties in the offspring. His first publication in 1905 identified a number of wheat characteristics that behaved as good Mendelian traits, including resistance to yellow rust. By 1910 he released Little Joss, a wheat variety resistant to this fungus.

It is also important to note that Biffen's work took place in a particular historical context in England. Britain had repealed the Corn Laws in 1846. These laws had placed tariffs on imported grain in order to protect English wheat farmers who had a hard time competing with the wheat imports of the Americas, Australia, and eastern Europe. Agitation by the working class and manufacturers led to the repeal, because the Corn Laws kept food more expensive than it would otherwise have been. Unfortunately for English farmers, once the protection of the Corn Laws was gone, they could no longer compete with foreign competitors. British agriculture entered a long, steady decline that left Britain dependent upon imported food.

Biffen quite consciously saw plant breeding science as a way to restore the glory of English wheat production, without re-instating the Corn Laws. In essence, Biffen sought to reverse decades of decay in the British agricultural economy. His motivation was only partly scientific. He also saw his work as an effort to correct something he thought was unnecessary (massive wheat imports) in the political economic sphere (Biffen 1905).

In his lifetime, Biffen was not successful in making the United Kingdom more self-sufficient in food. Nevertheless, through his service as the founding director of the Plant Breeding Institute, Biffen created an institutional legacy that kept the idea of self-sufficiency in food production alive.

Plant breeding science itself moved steadily ahead in both the United States and the United Kingdom. By the 1940s, plant breeders knew how to select parents to test for heritable traits, fix valuable traits in offspring, and separate heritable from environmental variation through statistical methods. Modern plant breeding methods also moved to many other places. In British India, for example, Albert Howard, another University of Cambridge graduate, was the Imperial Economic Botanist and bred wheat varieties that were still useful a half century later.

Without doubt, plant breeding by 1940 had contributed to increased and more stable yields in many crops. For all its practical importance, however, the science was still mostly a national affair in which breeders were supported by national research institutes and, in a few cases, private companies. This discipline was not a matter of foreign policy, and it played virtually no role in international politics.

Events during and in the aftermath of World War II changed this. Most importantly, the war produced victors over fascism, but they were divided themselves into two warring factions: the "free world," and the "Soviet bloc." The United States was clearly the leader of the first group as it managed the occupation of Japan and much of Germany and the rebuilding of Western Europe. At first the USSR was the leader of the second group, even though that leadership grew increasingly strained after the victory of communist forces created the People's Republic of China.

American political leaders, particularly President Harry S. Truman, embraced a comprehensive plan to contain and isolate the socialist states through military and other means. Included in the "other means" were the plant breeders. In ways that were unthinkable in the 1930s and before, plant breeders moved into the front line of the non-military dimensions of containing communism.

In Truman's inaugural speech in 1949, the fourth point (Point IV) was to create the means of sharing America's preeminent technological strength, particularly in agriculture, with less industrialized countries. This scheme to support agricultural research abroad, the "Point IV Program," had both political and humanitarian motivations. Poverty within the "third world countries" was real, and development theory at the time saw poverty and hunger as leading to susceptibility to communist subversion. Plant breeders, so the argument went, could assist in raising yields, alleviating poverty and hunger, and reducing the temptation to embrace the calls for communist revolution.

A program in Mexico started in 1943 by the Rockefeller Foundation provided the model for foreign assistance in agriculture through the Point IV Program. In the years after 1949, the US government, in collaboration with the Rockefeller and Ford Foundations, began a concerted effort to help countries like India modernize their agricultural research capacities and to urge national governments to embrace the promise of plenty provided by the plant breeders and other agricultural scientists.

At about the same time as these new institutional arrangements were being made, a wheat breeder from the US Department of Agriculture, Samuel Salmon, observed some unusual wheat plants being grown at research stations in occupied Japan. He arranged that samples be sent to the United States, and they included varieties that contained what were later called semi-dwarfing genes. Orville Vogel of USDA put this genetic material into varieties grown in Washington State by the 1960s. Similar successes by Norman Borlaug of the Rockefeller Foundation in Mexico led to releases of varieties for that country at the same time.

And so the "Green Revolution" began. At the request of Indian breeder M. S. Swaminathan, Borlaug arranged for varieties of the Mexican wheats to go to India, where they did well. More importantly, the Indian scientists at the Indian Agricultural Research Institute put the semi-dwarfing genes into varieties, some bred by Albert Howard decades earlier, where they did even better than the Mexican varieties.

What these new varieties accomplished was the efficient conversion of nitrogen fertilizer, plus ample water, into harvestable grain. Their straw was stiff and short, because the plant made grain, not straw. Moreover, the large heavy grainhead could stand up at the end of the season, because of the stiff, short straw. Earlier varieties of wheat tended to lodge or fall over when stimulated to excessive growth by nitrogen fertilizer. The Green Revolution varieties of wheat were veritable machines for turning fertilizer and water into grain, and their yields went up an astonishing two- to three-fold. Similar advances with semi-dwarfing genes in rice led to the same astronomical gains in that crop as well.

A final component of the Green Revolution must still be told. American assistance in plant breeding outside of the United States was, as mentioned above, driven partly by serious geopolitical concerns. Promotion of plant breeding science was, for the American government, promotion of peace through the containment of communism. Other countries, however, saw the situation differently.

For India, the question was not communism. India, after a century and a half of domination by the British and other European powers, did not want to have the Americans simply tell them what to do. At Independence in 1947, India developed a feisty nationalism guided by Jawaharlal Nehru, the first Prime Minister. Nehru wanted India independent of both the western and Soviet blocks. He wanted India to come into its own as an industrial power.

Nehru was strong-minded and led India to amazing changes until his death in 1964. He saw India create a more diverse economy including major industrial establishments. Nehru, however, was less interested in embracing the new agricultural science based on fertilizers and Mendelian plant breeding. He supported it fully in theory, but at budget time the rupees were scarce to support Indian agricultural scientists.

Nehru died in 1964 and was replaced by Lal Bahadur Shastri, who began immediately to alter agricultural policy in order to achieve greater domestic production. The new government took keen interest in the trial yields of the Mexican wheats imported in 1964. In 1965, outbreak of war with Pakistan and the beginning of two years of back-to-back drought (1965–1966 and 1966–1967) solidified the embrace between India's political rulers and its plant breeders. For the sake of India's autonomy and prosperity, said Shastri's government, India would seek new varieties and promote through price policies the production of more grain. India, in other words, would seek the promise of high yielding agriculture for its own national security interests. The Government of India may not have seen the world in the Cold War terms of the United States, but it certainly saw agricultural science as a way of securing Indian independence, security, pride, and dignity.

One other country's experience confirmed that nations sought high yielding agriculture through plant breeding in order to promote their own national interests.

The United Kingdom was left in 1945 as victorious but standing amidst the rubble of the Second World War. It was still powerful, but its imperial economy was in tatters and shortly to disintegrate with the independence of India and Pakistan. For over 70 years, Britain had relied upon imported food, particularly wheat as one of its main staples, but its economy could no longer support this way of living. Canadian and American farmers would not give their wheat to Britain, and Britain no longer had the foreign exchange to buy it.

In response to this crisis, the Labor government that came to power in 1945 took steps to resolve the problem raised years earlier by Rowland Biffen: can Britain breed wheat varieties that will make Britain self-sufficient? Clement Atlee, the Prime Minister, and his Minister of Agriculture, Tom Williams, presided over a multi-party agreement that created a new and vastly expanded Plant Breeding Institute. Through the Institute, British wheat breeders John Bingham and Francis G. H. Lupton eventually created new wheats based on the semi-dwarfing genes but adapted to British conditions. After the 1980s, British wheat yields were the highest in the world at around 7 t per hectare, and Britain was a net exporter of wheat.

The new state of British agriculture was not just a matter of science, however. Provision of extension advice to farmers, favorable prices for farmers, and, after 1972, Britain's entry to the European Community (now the European Union) were each important institutional pieces that supported plant breeding scientists. Like India and the United States, the government of the United Kingdom sought its national interests through the fruits of agricultural science. For their part, the scientists would never have received the support had vital national interests not been at stake. Roland Biffen's dream of using Mendel to make Britain self-sufficient in food had finally come to pass, but it had taken threats to national security to forge the political base needed for the scientists to do their work.

Agricultural scientists generally do not think of their work as a component of national security policy. Yet the advent of high yielding agriculture suggests that science for agriculture has close ties to the strength of nations and how national leaders may see the situation. Ethical issues abound in the realm of national security considerations, and scientists planning future work need to understand these connections.

Plant Biotechnology and Responses to the Criticisms of the Green Revolution, 1980–2000[8]

Yield enhancements achieved through the science of the Green Revolution reshaped the global agricultural economy, but they also left in their wake many criticisms. First, the new practices contributed to increasing social and economic inequality,

[8]This case study is adapted from a paper, John H. Perkins, Engineering the Earth and Changing Scientists: the Rockefeller Foundation and the Strategy of Agricultural Research, delivered to the Society for the History of Technology, Atlanta, Georgia, October 2003.

because poor farmers could often not take advantage of the technology. In addition, some argued that environmental damage resulted from increased uses of synthetic fertilizers and pesticides. Third, the new practices were potentially not sustainable, because of drawdowns in groundwater for irrigation and increased salinization of irrigated land. Finally and ironically, hunger persisted despite an increase in food supplies more than adequate to support every person on the planet.

Who, if anybody, would step forward to deal with these criticisms? Was there any institution to address the challenges as a part of its mission? How would agricultural scientists design their researches in the aftermath of introducing the high yielding varieties of wheat and rice? These questions were vexing, as answers were unclear. In 1980, agricultural science was still largely a national concern. Even with the international agricultural research centers and the Consultative Group on International Agricultural Research, governments had no common strategy for supporting agricultural science.

Was a coordinated, strategic plan for agricultural research needed to address the issues raised by the Green Revolution? This is a complicated question, because it presumes that scientific research could solve the problems created by the high yielding cereal grains. Yet many of the problems created by the new technology were social, political, economic, and cultural in nature, not scientific. More science and technology were not necessarily going to solve the important problems of inequality, sustainability, natural resource degradation, and hunger.

At the same time, it was clear in 1980 that new scientific research was needed. The Green Revolution, for example, had intensified the use of land by selecting genetic varieties that could respond to high inputs of fertilizer and water. Pest control was often needed to protect the increased production. If fertilizers, water use, and pesticides were the cause of important environmental problems, then perhaps science might create conditions in which the problematic materials (fertilizers, water, and pesticides) could be used more efficiently. Maybe alternatives to their uses could be found.

If scientific research had a role to play, then reliance on national efforts alone were unlikely to be adequate for the challenges. Countries in which the problems were most severe had the least capacity to build scientific programs. Additionally, rich countries with enormous scientific resources were not likely to put their scientists to work on the problems of poor countries. Most existing international institutions had the economic means to support research, but they didn't have the capacity to plan and coordinate it.

One institution, however, the Rockefeller Foundation, could plan strategically and implement coordinated international research programs. As noted above, Foundation staff developed this capacity in the early 1940s in Mexico.

For reasons that are yet unexplored, in the early 1980s the Foundation started a series of projects. They showed that, once again, the Rockefeller Foundation was one of and maybe the only major center for global strategic planning in agricultural research. As the successes of the Green Revolution ran their course and senior staff began to retire, Foundation officers convened an advisory group to review the needs of agricultural science. One possibility suggested was to create a new research program in "… the application of molecular biology to plant breeding" (O'Toole et al. 2001). This review led in 1984 to the launching of the International Program on Rice Biotechnology.

Over 17 years (1984–2000) the Foundation invested $104.7 million in the Program (ibid.). In the abstract the funds seem like a great deal of money, but it is important to recognize that they are quite small in relationship to other funds spent on agricultural research. In 1993, the highest spending year for the Program, expenses reached $8.8 million (8%), a small sum compared to the total amount of public and private funds invested in agricultural science research.

Despite the program's relatively low costs, its highly coordinated structure created a very active, productive research community with an associated education and training program. All of this, it should be noted, was aimed at remaking rice, the world's most important crop. Major lines of research included pest resistance, drought and flood tolerance, tolerance to saline soils, nutritional quality improvement, and comparative genomic studies.

One specific example of the Program's research illustrates its operations: the creation of rice varieties that make grains with β-carotene to help alleviate vitamin A deficiencies commonly found among poor people dependent upon rice. Ingo Potrykus and Peter Beyer at the Swiss Federal Institute of Technology in Zürich and the University of Freiburg in Germany, respectively, played pivotal roles. In 1999, they demonstrated that rice could be transformed to produce significant amounts of β-carotene, a chemical that the human body can make into Vitamin A. Vitamin A deficiencies cause a multitude of health problems, including blindness, and result in millions of preventable deaths each year (Sommer 2001). "Golden Rice," so called because of the yellow color of the β-carotene, had the potential to alleviate Vitamin A deficiencies among people with rice based diets.

Potrykus began work on plant transformation in 1970 when he worked on regenerating ornamental plants from protoplasts. He was aware, at the time, of the successes of the Green Revolution but felt that new methods would be needed for further scientific developments. Significantly, he believed that the expanding human population would require new knowledge (Potrykus 2001).

Rice, as Potrykus quickly discovered, was not an easy plant to manipulate for transformation experiments with DNA. Difficulties arose either in getting the new DNA into the rice plant cells or in regenerating whole new functional plants if the DNA was absorbed. Potrykus and his co-workers faced many disappointments and frustrations as they tried various ways of moving DNA into different kinds of cells and an assortment of methods to regenerate plants from cells. A major breakthrough came from a different lab's technique for treating a mass or callus of embryonic rice cells with exotic DNA and subsequently regenerating whole rice plants.

By the end of the 1980s, Swapan Datta working in Potrykus' lab used these embryogenic protoplasts to create the first transgenic rice of the Indica variety. Rather quickly they were able to produce transgenic rice that made the insecticide derived from the bacterium *Bacillus thuringiensis*. Only after Datta's successes did

Potrykus become interested in the prospect of producing transgenic rice with potential to alleviate deficiencies in Vitamin A.

Rockefeller Foundation staff had identified the problem of Vitamin A deficiency as top priority for the International Program for Rice Biotechnology at its inception in the mid-1980s (Toenniessen 2000). In 1992, Potrykus approached the Rockefeller Foundation Program for support of a new project to engineer a rice plant that could make β-carotene. By this time Potrykus, his colleague Peter Beyer from Germany, and others knew that rice needed four biochemical reactions in the rice seed to make β-carotene. This suggested that if they could transfer four genes into rice and make those genes function in the endosperm, the starchy part of the rice seed, they would have a plant that might alleviate Vitamin A deficiencies. Skepticism was high due to the difficulties of moving an entire, functioning biosynthetic pathway into rice. Other transformations that had previously succeeded involved inserting just one functioning gene (Potrykus 2001).

Despite the difficulties, Potrykus, Beyer, and a number of other colleagues succeeded within seven years. They had successfully moved three genes exotic to rice into the plant, and the three functioned to make the rice grain produce β-carotene. The public announcement of the creation of Golden Rice came at a conference in Switzerland in March, 1999, and formal publication was in *Science* in 2000 (Ye et al. 2000). Subequent work simplified the genetic transformation needed to produce β-carotene and increased its levels (Kathuria et al. 2007).

Despite the advances, commercialization of genetically engineered varieties from the Foundation's research program has barely begun. Nevertheless, the ethical implications of golden rice are hotly debated.[9] Coordinated, strategic research, however, leads to results, even if the lessons to be learned are not clear.

Consider the way in which the Rockefeller Foundation became the strategic planning center for agricultural research. It was largely by default both in the 1940s for what became the Green Revolution and in the 1980s with the rice biotechnology program. In other words, there was no other international, governmental, or non-governmental body with the resources, flexibility of mission, or staff that could have done this task. To be sure, if the fruits of research are universally popular, then lack of democratic procedure will not create ethical problems.

The Green Revolution, however, did create vigorous disagreements. One might also question if we can assume that genetically modified rice varieties will be free of criticisms? Other genetically modified organisms have attracted international controversy. Whether or not these criticisms reflect wisdom is beside the point. If the products of scientific research are controversial, part of the dispute may lie in the notion that parties affected were not consulted about the formation of the research programs.

To date genetically modified varieties like Golden Rice have not been incorporated into commercial varieties and widely adopted. Possibly they never will. Those

[9]For a strong opposition to Golden Rice, see Greenpeace (accessed 10 November 2007). An equally spirited defense of the transgenic golden rice was presented by Mayer (2005).

opposed have ethical concerns about transgenic crops. Despite opposition, some people may want to use the new varieties. This sort of controversy has occurred around the production of genetically modified maize, soybeans, canola, cotton, and other crops. Will this be the fate of the rice biotechnology program?

Questions of democracy, public participation, and ethics surround the science of applied biotechnology. Merely doing good science is not enough if a scientist wants to participate in the larger questions. This is a hard lesson to learn, but efforts to grapple with disagreements about scientific research are an important part of learning to understand the full complexity of new technologies.

Organic Agriculture: The Search for Qualitatively Different Intensification, 1930–2004[10]

Concurrent to the efforts being made to increase agricultural yields and efficiency from the 1940s through the 1970s, another sector of agriculture was maturing. Based on the principle that farming should mimic nature, organic was unique in seeking both biological and social intensification. Unlike any other sector of food production, organic farming eventually grew into a social movement. Buying an organic food became a political, a social, and an environmental statement.

Although he did not use the word "organic" to describe the farming method he promoted, Sir Albert Howard is often credited with founding many of organic farming's biological principles. As the Imperial Economic Botanist with the British controlled Government of India, Howard bred higher-yielding wheat varieties. After retirement from government service, he authored *An Agricultural Testament* (1940), which formally introduced the Indore Process, known today as composting.

He was wary of the effects in soil of mono-crop, intensive agriculture. "The capital of nations which is real, permanent, and independent of everything except a market for the products of farming, is soil" (ibid.). Howard suggested that to maintain the integrity of soil, farmers must compost, cover-crop, rotate crops, all fundamental features of contemporary organic farming.

Lady Eve Balfour built on Howard's work by initiating the first and longest comparative study of the microbial impacts of organic compared to conventional farming methods. The Haughley Experiment, conducted through 1972, sought to:

> Observe and study nutrition cycles, functioning as a whole, under contrasting methods of land use, but on the same soil and under the same management, the purpose being to assess what effect, if any, the different soil treatments had on the biological quality of the produce grown thereon, including its nutritive value as revealed through its animal consumers.
>
> (Balfour 1950)

[10]This case study was adapted from Jamison (2003).

This study was the first to suggest that microbial activities and absorbable nutrient levels on organic farms were comparable to those on conventional farms. It also made clear that pest pressures were similar on organic sites.

Balfour integrated human health into her arguments for organic farming. Her work led to a social outcry for food that was produced without synthetic chemicals. "My subject is food, which concerns everyone; it is health, which concerns everyone; it is the soil, which concerns everyone – though they may not realize it."

Nearly 20 years later, in 1962, Rachel Carson's *Silent Spring* confirmed and enhanced what Balfour had suspected about pesticides and human health. From cancer to infertility, Carson focussed on the deleterious effects of synthetic pesticides and other persistent chemicals. She also highlighted damages to wildlife as an additional problem. Many credit her work with the ultimate banning of DDT, the establishment of the Environmental Protection Agency, and the growth of the organic food movement.

The organic food movement, strongly associated with the back-to-the-land movement, was like few others. The movement united environmentalists with human health advocates. It challenged companies selling synthetic pesticides and fertilizers. It contested the idea that humans could control natural systems. Finally, it instigated one of the fastest growing agricultural markets of the late 20th century.

Fundamental to the philosophy underlying the early organic food movement were the ideas that "organic" supported local food systems, small farmers, and the environment. Organic farming was more than growing food, it was a way of life. As its popularity grew, organic food stood for safety.

The Alar incident in the late 1980s demonstrated the immediate response on the part of consumers to the perceived threats posed by synthetic agricultural chemicals. Alar is a post-harvest material used on apples to preserve crispness and appearance from field to market. In 1989, the television program *60 Minutes* reported that residues of the chemical posed health threats to children by increasing their risks of developing cancer by up to 100 times. Apple sales immediately plummeted. The incident proved that consumers responded to issues of food safety and would pay higher prices for organic food.

Today, organic food is the fastest growing agricultural sector in the United States (USDA Economic Research Service 2003). In response to its' rapid growth, the United States Department of Agriculture implemented the National Organic Program (NOP) in October of 2001. The NOP set the standard to which all food sold in the United States as "organic" must be produced.

The current rapid growth in sales of organic products has come increasingly from industrial-scale organic farms. To many early advocates of organic farming, the robustness of the industrial sector threatens the people whose work gave rise to the organic market. Farmers who participate for environmental and social reasons will face challenges similar to those faced by small farms as conventional agriculture industrialized (Strange 1988).

Organic has long been regarded as the philosophical and scientific counterpart to conventional agriculture. It attempted to work with ecosystems instead of controlling them. It promoted rural communities and local food systems as opposed to global markets.

Organic agriculture currently rests at a crossroads. One direction maintains a strong connection with its roots in local, diverse food systems. The other leads to an industrialized organic sector. Consumers will ultimately choose the road.

Agricultural scientists, whether they are involved with research on organic production or not, will be influenced by the science of organic agriculture and its ethical principles. The organic idea provides an alternative for agricultural research.

Conclusions

Agriculture has had rich and complex effects on human life and culture since the Neolithic age. Most of those effects resulted from the increasing intensification of resource use, particularly of land and labor. Professional agricultural scientists, supported by public and private funds, were pivotal in the intensification processes. From the case studies above, four general themes unify the ethical concerns facing future work in the agricultural sciences.

First, changes in agricultural practices affect not only the scientist and the farmer but also the entire human population. Scientists and farmers may be the most intimately involved, and frequently they are the groups most influential in the support of agricultural research. It is important to realize, however, that every person on the planet is affected and has a vital interest in the outcomes of scientific and technological developments. When people are not involved in decisions that may profoundly affect them, the ethical issue of "prior informed consent" is present.

For example, the development of synthetic pesticides conferred many benefits from intensified use of land, but it also created severe contamination. No person gave informed consent to the uses of technology such as DDT that left residues of toxic material in their tissues. Integrated Pest Management (IPM) and organic agriculture are now both correctives to problems involving synthetic pesticides. Decisions on their uses, however, were left largely to farmers operating under constraints of highly competitive markets. Again, little information was made available to help citizens understand the tradeoffs involved.

Similarly, developments in classical plant breeding and in breeding based on molecular biology resulted in higher yields and the potential for other benefits in nutrition and plant hardiness. Both the Green Revolution and the Rockefeller Foundation's program in rice biotechnology, however, were strategic programs of scientific research that had no prior informed consent from either individual farmers or general citizens. Does the lack of prior informed consent make the practices unethical? Not necessarily, but the issue still remains.

Even the science behind organic agriculture, often presumed to be entirely ethical by its advocates, cannot escape from demands of prior informed consent. Sir Albert Howard, the British wheat breeder who in the latter part of his career laid part of the scientific basis for management of soil tilth through organic farming methods, made his observations about the Indore process among peasant farmers who were not entirely free. Britain at the time exercised imperial control over India, a fact that affected the entire range of social and economic features of the country. This was science that not only lacked prior informed consent from those affected, it also had no prior informed consent from those participating in the studies.

The second major theme emerging from these case studies is that the results of agricultural science have profound ethical effects through influences on political power. Two different levels of power are involved: (1) the control an individual farmer has over his/her own land, and (2) the aggregate power that a nation state derives from the productivity of its land.

Control of land by an individual farmer is intimately connected to yields and their values. In general, an individual farmer who obtains higher yields will more easily maintain power over his/her land. This situation is especially important in market economies based on commodity trade and markets in land. As market economies became the norm during the 19th and 20th centuries, farmers who could not use science to intensify their uses of resources were increasingly disadvantaged.

Another way of viewing this problem is to note that farmers had unequal access to agricultural science. Based on age, education, gender, race, health, and many other factors, some farmers were in better positions to adopt the new practices emerging from the agricultural experiment stations. These were the technologically progressive farmers who expanded the scales of their operations at the cost of other farmers who could not compete and thus often lost control of their land.

Economist Willard Cochrane has explained this phenomenon as the "treadmill hypothesis," in which the early and most skillful adopters of new technology supplant the late, never, or unskilled adopters (Cochrane 1993). In a process now seen across the world, farmers became fewer in number and larger, and many people left the countryside for life in the cities. Sometimes leaving agriculture was voluntary and sometimes not. Sometimes that life was better and sometimes not. No matter the circumstances, the rural to urban transition, based ultimately on the work of agricultural scientists, was ethically charged.

The second form of political power involved focuses on the aggregate power accruing to nation states from the high yields made possible by intensification. A country that is not self-sufficient in agricultural production must obtain needed produce from other countries. As a result, the country is vulnerable. A country with a surplus of yield, in contrast, will be more stable. In short, surplus confers power internationally.

This question of national power was most evident in the agricultural science that created the Green Revolution. As noted above, the United States, India, and Britain each embraced the science of high yielding wheat and rice in order to secure what their respective governments saw as national security and power. Exercise of that power, of course, always involved ethical considerations. Individual agricultural scientists cannot by themselves resolve these problems. It is not unreasonable, however, that they understand that their work will have geopolitical effects.

A third theme centers on the connections between agricultural science and human health. Traditional boundaries between sciences have usually sharply separated agricultural scientists from biomedical and health scientists. They have little interaction professionally or institutionally. Governments place the concerns of agriculture and health/medicine in different departments and seldom search for connections.

Agriculture, however, is the foundation for health. Malnutrition from low and unstable yields is a predictor of health problems. Yield enhancement and stabilization from synthetic pesticides and IPM can alleviate malnutrition. Enhanced yields from Green Revolution grains and biotechnology can improve diets. Organic produce with fewer pesticide residues may also have health benefits.

Whether or not enhanced, stabilized yields actually help nutritional status, however, is also important and depends upon the distribution of the harvest. Specifically, a person's access to food generally depends upon either control of land for subsistence purposes or an adequate income. Because of the treadmill effects noted above, new agricultural practices can impact distribution. If a person loses land for subsistence and has an inadequate income, yield increases don't protect health. Thus the results of agricultural science can sometimes harm nutrition and health.

The final theme is the effect agricultural science has on the global environment. Resources devoted to agriculture are enormous. For example, about one-third of the soil resources of the earth are in agriculture, and 90 percent of the water under human management is used in agriculture. Humans harvest approximately two-thirds of the net primary productivity of photosynthesis on earth, and this proportion may grow with increases in population.

Again a situation with dual consequences emerges. On the one hand, human beings must utilize agriculture to survive. The intensification involved of the Green Revolution has been key in producing enough calories to potentially feed 6.3 billion people. Similarly, synthetic pesticides, IPM, and plant biotechnology may have irreplaceable roles to play as population continues to climb.

Unfortunately, no matter how much people depend on these technologies, some of them have potentially disastrous consequences. Mobilization of water resources for irrigation can salinize soils, deplete ground water, and destroy biodiversity in rivers, lakes, and oceans. Synthetic nitrogen fertilizers, based on immense energy consumption, are polluting water and releasing greenhouse gases that threaten the climate. Energy supplies may become increasingly expensive or impossible to use due to climate change. Some uses of synthetic pesticides have had deleterious impacts on the health of farmers, farm laborers, consumers, and, especially, children. Transgenic plants may erode genetic and ecosystem resources in still unforeseen ways.

Agricultural scientists are thus faced with monumental challenges: how to maintain stable yields without jeopardizing agriculture or the earth's resources. Two case studies, on agroecology/IPM and on organic agriculture, demonstrate routes to intensification that may be more environmentally benign.

Agricultural science and agriculture pose numerous complex ethical dilemmas. Agricultural scientists need to decide research priorities and methods with the certain knowledge their research will have ethical effects. Issues of prior informed consent, individual and national political power, human health, and the environment complicate the choices that scientists and their supporters must make. The future prosperity of people and all other species depend upon the choices made.

Chapter 4
One Hundred Years of Agricultural Intensification: A Personal History of Unanswered Ethical Issues – 1890–2004

A. Allan Schmid*[1]

Perhaps I can best add to John Perkins' historical overview with a micro example of intensification. I never thought while growing up on a Nebraska farm that my family's decisions raised any ethical issues. I never thought we were exercising power over others. I never gave much thought to the fact that our choices were made within a larger framework given by society. This conference has made me think again. We were ethically responsible for our individual choices within this framework as well as in our capacity as citizens to examine and critique this structure of rights. A series of management choices made by my family over the 100-plus year period 1890–2004 will be examined for their broader implications for other people, which I take is part of what constitutes ethical issues. The ethics embedded in the informal social and formal political framework that structured these management choices will also be examined. I believe it is consistent with Perkin's "political ecology" approach linking ecology, economics, and politics.

Credit Source

Intensification requires credit. My family farm is at the western edge of the U.S. corn belt, which means that rainfall is uncertain. Cash flow is also subject to price shocks. This means that if credit sources do not take the long view, sooner or later loans can't be repaid with this year's crop or maybe even next year. My maternal grandfather lost his farm in the Great Depression and spent his old age as a grocery clerk, but thereby qualified for Social Security benefits at a time when farmers were not eligible. My parents kept their farm business going, in part, by some years of rent deferral from my father's mother who owned the farm. Parents have a sympathetic morality not common to commercial banks. My father reduced his dependence on borrowed capital by sharing the purchase of machinery with his brother. Through cooperation based on

---

* University Distinguished Professor Emeritus in the Department of Agricultural, Food, and Resource Economics at Michigan State University.

[1] Thanks to Larry Busch, Steve Esquith, and Paul Thompson for helpful comments.

P.B. Thompson (ed.) *The Ethics of Intensification*,

familial trust, they could pool their savings and avoid going to banks. My grandfather had bought his land with a loan from a relative who had immigrated from Switzerland before him. Who knows if the relative would have foreclosed if my grandfather had had a cash flow problem from weather or price shocks?

In the public sector, historically the Farmer's Home Administration (now Farm Service Agency) of the U.S. Department of Agriculture was more accommodating to short-run cash flow problems in repaying farm real estate loans than the private commercial loan sector. Whether this represented an ethical judgment and sympathy for distressed farmers by the general taxpayer or just the political power of farmers is hard to know.

Agricultural Adjustment Act of 1933

"You mean we are going to kill little pigs to improve farmer incomes?" During the Great Depression the concern was not with animal rights, but the incongruity between people going hungry alongside agricultural surpluses that depressed farm incomes. The broader society made a moral commitment to farmers (and others) to help them survive the Great Depression with subsidies and facilitation of collective action to manage supply. This macro policy helped my parents survive.

Seeds

Intensification requires new seeds embedded in more research and development. The use of seed raises several ethical issues. Soybeans were introduced in the U.S. in the early 1900s. We did not compensate Asian farmers for the germplasm. The issues of who owns the world's genetic heritage continues to be a source of conflict. The nitrogen fixation capability of soybeans greatly increased productivity of our farm since 1940.

My parents bought a wheat variety and then saved their own seed until its vitality eroded. Was this a kind of theft? This limited the recovery of private research and development costs – though for years research was primarily publicly financed in agricultural research stations. In the case of maize, hybrids stopped farmers from saving their own seed and required them to return to the seed companies each season.

Ethical issues are involved in whether property rights should apply to seeds at all as well as whether farmers should have the right to replant purchased seeds.

Grade A Milk

My grandfather, with his Swiss heritage, had a few cows and made his own cheese. My parents' survival strategy included diversification. They kept a few cows and sold cream, hauling it to town once a week in the 1940s. I can remember when they sold whole milk for a period, and the metal cans stood at the side of the road in the sun until the dairy truck came to pick them up. I suspect there were some health risks involved. When states passed new requirements for production of Grade A milk, farmers needed to make new investments in milking machines, bulk tanks and coolers. This put an end to small farmers having a sideline in milk. You either got big or got out. The new requirements, no doubt, improved sanitation & health, but destroyed some fixed immobile assets. The barn for the cows was now worthless and unsuited for other uses. The beneficiaries did not compensate small farmers for their losses. Intensification led to specialization and the need for more capital.

An aside: The impact of higher sanitary rules on old style farming is still going on as EU rules are applied to new members such as Poland. The EU requires septic tanks for cow waste and separate facilities for hogs and cows. These intensification requirements will be impossible for many Polish farmers to meet.

Diversification included raising chickens and selling eggs. Baby chicks were purchased, broilers dressed primarily for home consumption, and eggs sold in town. Now economies of scale are such that only large specialized broiler and egg production are viable. Every farmer in our community of cash grain farms buys milk, chicken and eggs at the grocery store like everyone else. Farmers have lost the advantage of diversification and consumers have gained lower prices from specialization.

Labor

The organization of labor raises a number of ethical issues. Tasks requiring more than the labor of the farm family can be organized cooperatively or by hired labor, including migrants. The pre-1940s threshing crew of neighbors avoided the need for seasonal migrant workers. The community also provided disability insurance. If a farmer had an accident or health problem preventing them from planting or harvesting, the neighbors would do it without compensation. Child labor was, of course, common. Was it abused? No one worried about it. My father quit school after the eighth grade.

The central livestock auction yards, in large cities in the Midwest such as Chicago and Omaha, are now gone. Slaughterhouses escaped big city unions and moved to small towns using cheap, often immigrant, labor.

Soil Conservation

Management of soils has always raised the moral issue of what is the responsibility of the present for future generations. Does a farmer forgo short-term profit to invest in terraces and other conservation practices. Not all farmers made the same judgment. My father used crop rotations with legumes rather than continuous maize at some cost to immediate returns. During his lifetime, he terraced the entire farm connected to grass waterways that reduced tillable acres in exchange for long-term sustainability. Erosion and ditches were noticeably reduced. His stewardship was made easier by government programs of cost-sharing. Society accepted some of the cost for its long-term welfare.

Inheritance

The mode of inheritance raises ethical questions within the family with implications for the community. If all siblings share inheritance of the land equally, there is either fragmentation or land is sold outside of the family. The outcome is related to family size. My father was the youngest child and stayed on the home farm helping his father. He saved some money and helped one of his older brothers buy a farm. When my grandfather died, my father rented the farm from his mother as noted above in the section on credit. When she died, he bought out his five siblings. Today, it would be difficult to accumulate enough savings to avoid a mortgage to a bank.

Endangered Species

I witnessed the disappearance of long-eared jackrabbits (lepus). Intensification can destroy habitat and greatly reduced the numbers of ground hogs, coyotes, and native prairie grasses. I never gave it a thought. In the case of nasty ground hogs and chicken-killing coyotes, good riddance I say. Some environmentalists may think otherwise. Whose animals are these anyway? There was a period when the state paid us a bounty for every coyote killed. On the other hand, today with government incentives, some native grasses are being reintroduced. The government pays us to keep erodable land in grass, but I derive a certain satisfaction from seeing Big Bluestem once again, towering above the heads of my grandchildren.

Disease Control

My father fed some antibiotics to cattle for growth enhancement, one example of intensification. The use of anti-biotics is even more important today as the animals in large animal "factories" are more stressed than the 50-head operation of small farmers. Previously diversification of crops and livestock minimized disease buildup compared to today's specialized animal feeding. Farmers with a moral commitment to the welfare of their animals and the humans eating them complied with laws requiring the cessation of additives some time before slaughter. A corporate owner may be less conscientious and identify less with consumers. It is the difference between a sense of feeding the world and just moving units of product. Some can't imagine that farmers have such a world view, but my personal experience suggests otherwise.

Herbicides

My grandfather initially bought 160 acres, which was a family farm in eastern Nebraska. My father bought an additional 160 acres and it was still a family operation. My cousin who rented our land upon my father's retirement, farmed his father's 320, plus 320 of his wife's family, plus my father's 320, and rented more besides. It was still a family operation with modest income and no hired labor. Together with larger tractors, herbicides were responsible for one farm family to go from 160, to 320, to 960 acres in three generations – sprayers cover more ground per hour of labor than mechanical cultivation. I know I plowed out some maize driving too fast. Neither my cousin nor myself gave much thought to possible ground water contamination and chemical residues.

Pesticides

Remember our degree of self-sufficiency and diversification. I recall using Paris Green (copper oxide) to control pests on our fruit trees. Upon application, I was green as well as the trees. I'm sure it was not healthy. I remember when we first used DDT to control flies on the cattle. We thought it a miracle as the flies died and the animals did not have to waste energy swatting them with their tails. Little did we know that the flies and other pests would evolve requiring even more chemicals. Many crop pests were minimized by rotations and diversification that is no longer profitable.

Energy

A large barn on our farm was built to house five teams of draft horses plus carriage horses. The barn was the support system for the horses with hay above and a bin for oats. Renewable oats fueled horses rather than non-renewable petroleum for tractors. The implications for the geo-politics of oil are obvious. The impact on agricultural output is also obvious in retrospect. Land formerly used for production of oats for horses now could be used for crops to feed animals for human consumption. Not all of the increased production from agriculture is due to better genetics and management, but from substituting Texas and Middle East oil for homegrown oats and hay. The substitution of the car for the horse-drawn buggy should also be mentioned.

Culture

One result of larger and fewer farms was the loss of rural schools, churches, and towns. A unique culture went with them. Many of the remaining small towns in the Midwest and Great Plains are filled with old people and are not sustainable. Does anyone care? And if they did, what to do about it is not clear.

Conclusion

Efficiency depends on whose interests are a cost of production. There is no such thing as "true cost." Alpha's interest is no more true than Beta's. All accounted costs are selective human artifacts. Farmers respond to shifting prices, and prices are a function of rights. If you change who is buyer and who is seller of an opportunity, you will change price (willingness to buy and sell) of that good or opportunity to use it. Change rights and associated income distribution and you change prices and therefore the efficient reaction of managers to relative prices. Efficiency is not a unique outcome. There are as many efficient outcomes as there are different distribution of rights. Distribution of rights is a matter of ethical judgment. Efficiency cannot be a guide to ethics, rather a particular efficiency is an outcome of ethics implied in rights.

For example, large hog factories moved from traditional regions like Iowa where maize is grown to North Carolina where labor was cheaper as farmers could not make it growing tobacco were looking for jobs. Is this efficient? It depends on whether damage to estuaries is made a cost of production. Someone's interests will be foregone no matter what is the distribution of rights. Whose forgone opportunities are a cost to the other interdependent parties is an ethical question. A few years ago, a hurricane caused waste in animal holding ponds to overflow and enter the rivers and then be flushed to estuaries on the Atlantic coast. Should the subsequent fish and wildlife kill be a cost of hog production? The answer includes an ethical judgment about the use of power. As noted above, the same issues arise in the use of credit, seeds, product sanitation, labor, soil conservation, inheritance processes, endangered species, disease and pest control, energy use, and cultural change.

Utilitarian and rights-based arguments are not independent of each other. There is no way to sum utility without a foundation of rights, whether they be formal or informal, deliberative or not. The argument here puts ethics front and center. The moral issue is choosing (or accepting without circumspection) the system of rights and exposures. The fact that many of us respond to relative prices and available opportunities does not mean that our choices are not affecting others. It does not mean that we have no responsibility to forgo some given opportunities or to participate in the debate over the rights structure giving us our opportunities. It is the job of scholars and others to remind people of their interdependence and ethical responsibilities. I am only a little sorry if it makes life more complicated.

Chapter 5
Two Battles in the History of Agriculture: Against Hunger and Against Alternatives. Comment on John Perkins' and Rachael Jamison's "History, Ethics and Intensification in Agriculture"

Michiel Korthals*

One or Two Battles?

History of agriculture is often seen as a constant battle against nature in trying to end famine and hunger and John Perkins agrees with this view. The evil here, the enemy to be battled against so to speak, is the constant threat of hunger and under-nourishment. This battle, as all battles, has its proponents and its opponents. Proponents are in favour of more and higher yields, opponents are the critics of high yielding agriculture and intensification. Seen from this perspective, ending famine and hunger is the main ethical problem, and critics of the current approach to this problem are reproached to delay the indeed authentic efforts to end this evil. From this point of view, agriculture is mainly an issue of farmers that are willing to intensify or not, and since the 20th century they are increasingly assisted by professional agricultural scientists, which indeed resulted in the doubling and tripling of yields, as Perkins indicates.

The Battle Against Hunger as an Ethical Task

Hunger is seen by nearly all as an evil. The fact that a lot of food in the world is not fair or equally distributed, either because of lack of producing or because of lack of access and management, is widely considered as one of the most shameful things in the world. Ethically seen hunger is a phenomenon that is rather easy to identify: the food is simply not equally and fairly divided, which means that principles of equality and fairness are distorted with the result that some are fat and others are thin, under-nourished or even starving. Although many different interpretations of the principles of justice and fairness exist, identifying these kinds of misdistribution is relatively easy because the appeal to these principles have something commonsensical.

* Professor and Chair of the Applied Philosophy group at Wageningen University and Research Centre.

Perkins excellently shows that during the last four decades it has become clearer that consumers have a lot to say on the type of agriculture and food production, and have some influence on the course of agriculture taken. The first clear case of consumer/citizen intervention in agriculture and its science happened after the publication of *Silent Spring*, the book of Rachel Carson. "The crisis of pesticides was real, and legislation enshrined environmental protection as the driver of pesticide regulations." This interference had a considerable impact on science: "Agricultural scientists today, however, have moved to see the problem of pest control in a vastly different light." The second example Perkins gives is plant breeding and its geopolitical relevance. National interests were behind the drive for high yielding variants after the Second World War. The third example is that of biotechnology, in particular of rice, the discontent this aroused in many countries, and the concrete management implications. The examples given bring Perkins to the conclusion that the wider public cannot be left out with respect to important agricultural innovations.

The Second Battle: Against Alternative Agricultural and Food Styles

However, agriculture is not only about higher yields, but also about the type of foodstuffs to be produced, distributed, sold, bought, prepared and consumed, and it is this alternative battle that Perkins and Jamison neglect fully. Food is intrinsically connected with socio-cultural identities people construct, and a selection of certain food stuffs functions as identifier vis-à-vis other groups. Food as connected with the life style of a group, communality or nation, is value laden: it functions in distinguishing this social entity from other social entities and it supports this social identity. This process of distinction is often not peaceful but violent, and accompanied by power, be it war, pressure or other forms of coercion towards the groups that don't want to comply with a certain farming and consumption style. The determination of the type of food production through constant battles between dominant and less dominant parties is therefore also a prominent feature of the history of agriculture.

In my view, the historian should not only look to the history of agriculture from the angle of the battle against hunger but also from the angle of the battle between types of agriculture and the way a dominant type of agriculture is used as a tool in structuring society, both culturally and politically. From this point of view, consumers have more input in the history of agriculture, be it negative or positive, than from the first point of view, because in the end it is their preferences and acceptance of the dominant form of agriculture that count. Now, due to complex historical and social factors, consumers were during a long time of history identical with farmers. This means that the common preferences of consumers and farmers could be felt in the roads taken or not taken to improve or intensify agriculture. However, in the 20th century, consumers became more and more separated from farming, or better, separated from food production at large, which means that since that period it takes time before their voice is heard in agriculture. But it doesn't mean that there were no battles on the style of farming, producing, distributing, preparing and cooking foodstuffs, as Perkins shows. The last battle, still going on, is that on genetic modification of crops. In particular, in Europe this battle is heated, because genetically modified food is seen as part of the larger conglomerate of industrial farming against the cultural identity of many communities.

Fair Representation of Food Choices as an Ethical Principle

In contrast to the battle against hunger, the battle with respect to fair representation of food choices is not appealing to principles of just distribution, but to principles of respect for food choices and of voice. According to the first type of principles, equal and fair distribution of goods (food) is ethically justified because food is an important instrument for survival, or, better, living the life you want to live. According to the second type of principles, fair presentation is ethically justified because food is an intrinsic part of living a good life that persons mostly collectively choose to live. These kinds of principles govern *firstly respect for the food choices*, which implies respecting them even when they look from an other point of view on the good life as unhealthy or irrational. As a matter of fact, constraints should regulate what ideal of good life is ethically permissible and which not, but that is not my point here. The *second part of these principles should cover principles of voice*, in the sense of enabling the respected food choice indeed to be realized in producing, preparing and consuming food. These principles or rights can be seen as principles of (fair) representation of the respective food choices in production and consumption. Collective food choices should be heard in the production chain and on the market by the establishment of farming and food styles that correspond to these styles.

Fair representation of food choices can be justified both in the deontological theory of Kant as in the utilitarian theory of John Stuart Mill. The German philosopher Kant states:

> Laziness and cowardice are the reasons why such a large part of humanity, even long after nature has liberated it from foreign control (naturaliter maiorennes), is still happy to remain infantile during its entire life, making it so easy for others to act as its keeper. It is so easy to be infantile. If I have a book that is wisdom for me, a therapist or preacher who serves as my conscience, a doctor who prescribes my diet, then I do not need to worry about these myself. I do not need to think, as long as I am willing to pay.
>
> (Kant 1785)

As it belongs to ones autonomy, consumers should decide about their own food (diet); as a consequence, markets should follow. As is clear from Kants quotation, he presupposes that an adult is *educated*, has *capabilities*, and gets the (reliable) *information* on the diets that he or she wants to life with. Moreover, it presupposes also that indeed the production processes and markets deliver the goods and services an autonomous person prefers. Ideally, fair representation would mean that all food choices are treated equally in farming and production and find their expression in corresponding farming and production styles.

However, also from a *utilitarian* perspective, fair representation can be justified, although in a different way, as is clear from John Stuart Mill's statement: "His own good, either physical or moral is not a sufficient warrant. The only purpose for which power can rightfully be exercised over any member of a civilized community, against his will, is to prevent harm to others" (Mill 2000). Again, from this perspective, the autonomous person should be enabled to strive for his own good and food through education, regulation, reliable information and responsive markets.

As said, fair representation is a contested concept, and the history of agriculture is a continuous battle fought between parties that were already represented in agriculture and others that weren't or faced exclusion. Fair representation is an ideal, and more often than not, misrepresentation is the outcome of these battles. The various bones of contentions of these battles reflect the three aspects of representation. First, food products of the various food styles must be available and the production process must be so structured that it indeed these products are produced. Secondly, the food stuffs must also be affordable, i.e., consumers' prices must be more or less comparable without too much differences between basic food stuffs. Finally, food stuffs must be enjoyable culturally unrepressed, without constraints. People should be able to choose, e.g., a vegetarian food style without being ostracized or in another way punished because of their life style.

Implications for the History of Agriculture

Seen from the perspective of battles for representation, history of agricultural intensification is not only a battlefield for higher yields, but also for farming and production styles, that compete for dominance or, if possible, peaceful coexistence. The history of agricultural intensification is a complex network of roads taken or only half way taken, of crucial decisions on crossings between different farming and food styles.

Therefore, not only the last decades have these battles between diverging types of agriculture been fought. In fact they were constantly intertwined with the battle against nature for higher yields. Protests against displacement of labour are to be seen as phenomena that fit into the perspective of the battle for representation, and they are well known events in the history of agriculture. They did not only happen with machines or pesticide and herbicide resistant variants, but also when sheep or specific crops were used.

Seen from the possibility of misrepresentation, power does not only play a role in determining farmers to have access or not access to preferred seeds, resources, technologies, but also in preventing consumers to have access to their preferred food stuffs. There are several possibilities here: consumers are supplied with food stuffs, but not their preferred ones, and they refuse to consume. This doesn't happen very often, because under most circumstances people are willing to survive and to feed their kids with whatever is supplied. What happens more often is that consumers (and farmers) grudgingly accept the food, but feel denigrated and disrespected or instrumentalized. Consumers and farmers are trapped, because they have to consume and produce anyhow, so present them resources without choice, and in the end they will have to take these anyhow.

In the table below I enlisted the various features of both battles, by distinguishing between the way the respective evils are socially structured, ideological shaped and politically fought (in particular by governments). With respect to the power involved, the control farmers have over their land is in the case of misdistribution in a different way implied than in the case of misrepresentation. In the first case the farmers don't have access at all to seeds and other resources. In the second case, the farmers don't get the resources they prefer and they are under pressure to accept the seed and other resources that are offered. Kloppenburg's nice history of hybrid

Features of battle	Misdistribution of food	Misrepresentation of food styles
Social evil	Hunger, malnutrition	Repression of food styles
Ethical evil	Injustice (no equal distribution)	Disrespect (against sovereignty, voice)
Social abuse	Food as power instrument: blackmailing with hunger	Food, agriculture as instrument of repression: disrespecting
Moral ideologies of abuse	Hungry don't deserve food, their own fault	Disrespected food style is unhealthy, risky, harmful
Power object	Farmers	Consumers
Government: inside	Food as instrument against inner opposition	Politics against unwilling minorities
Government: outside	Geopolitics against different social systems	Geopolitics against different food cultures

maize fits into the second battle. He makes it clear that the financial pressures on farmers to accept the industrial produced hybrid maize seeds were considerable, even though they lost the possibility to reuse their own seed. What he also shows, is that at the same time there were alternative roads that in the end were not taken because of the pressure from that same industrial and governmental actors (Kloppenburg 2004). U.S. Secretary of Agriculture in the 1930s was Henry A. Wallace, who also at that time founded *Pioneer Hi-Bred*, the company that produced the seeds. So, in this case, not misdistribution was at stake but something like the lack of representation of alternatives to hybrid maize.

For the consumers the two cases are also differently structured: in the case of hunger, they don't get (enough) food by lack of purchasing power or availability, be it through lack of production or distribution. In the second case of misrepresentation, the supply of food can be sufficient, but is not accepted due other preferences, or accepted where by the negative implications (like obesity or food related diseases) are only later discovered. In fact, in most cases, acts of misrepresentation are directed towards consumers; in the end, they will have to accept the food stuffs even if these don't meet their food choices.

Conclusion

In this short comment on the historical narrative of Perkins and Janison, I tried to show that the perspective of the authors covers only the battle of mankind against hunger and not the battles between the different food styles for presentation in production and consumption. I made a plea for considering misrepresentation as an important evil in the food and agricultural sector that is very much entwined with the other evil, hunger, but has also some features of its own, like lack of pluralism and of representation with respect to food choices. In contrast to misdistribution of food which implies often fat bellies for some and hungry bellies for most, lack of pluralism is a lot more difficult to identify, although its outcome can as be disastrous and shameful as the first evil. The outcome is always something like misrepresentation, which means in the food sector that not all food styles are represented on the market, in the production, in governments and or in research. Bluntly said, it means that some groups have no voice in the food sector and some have more (disproportionate) voice.

This presupposes that the right on food choice of collectives or individuals is not respected. When, for example, the cheapest and most accessible type of food is fast food, and people at home are encouraged to eat as much as possible processed food and to do the least themselves, when in the production of food the only target is more yields, regardless of the quality of the food, of the environment, of the landscape, of the animals involved, than indeed, only one style is dominant, suppresses the others and there is an indication of misrepresentation (lack of pluralism).

Lack of pluralism is undeniable connected with the controversial concept of *quality of food*. This multi-interpretable concept is defined by different cultures in various ways. Quality of food is for a Moslem different for a Hindu or a Jewish person, to name only the most well-known and broadest life style groups. Although quality of food is controversial, some general aspects can be discerned which makes it possible to compare the different food styles. These aspects are for example the measure of respect for the animals involved, the landscape, and the environment. The different food styles differ with respect to these aspects.

Chapter 6
Agriculture Intensification from the Perspective of Development Ethics

Luis Camacho*

Even at first glance it is obvious that many of the problems making headlines today are related to agriculture. To mention just a few of them:

- Loss of self-sufficiency in grain production in many countries
- Need for strategies to cope with the falling prices of commodities like coffee
- Decreasing percentage of the price of the finished product sold abroad that reaches the farmer at home
- Political instability partially caused by failure of imposed crop substitution in South American countries like Bolivia
- Losses in agricultural jobs in Mexico after the signing of the North America Free Trade Agreement
- Conflicts of interest between consumers and producers in developing nations in connection with agricultural subsidies in Europe and the United States
- Passivity of agricultural sectors when faced with structural adjustment plans
- Loss of biodiversity due to monoculture and use of pesticides
- Elimination of subsidies for farmers, especially in negotiations of free trade agreements

For most people, especially politicians and government officials, these problems merely call for technical and political solutions. Discussions on the criteria for the decisions being taken are usually pushed to the sides. Normative questions are seldom brought into focus when problems are analyzed and solutions proposed, probably because it is difficult to find consensus in the method to be followed. Moreover, a shallow approach is always tempting and frequently pursued: when ethical principles are sought for the analysis and proposed solutions of these situations, one is likely to invoke in a hurry such notions as distributive justice, justice as fairness, equality in rights, maximum possible benefit for the majority and least harm for all, participation of individuals in decisions affecting them, and a few similar other principles and norms. But there is room for a more careful application of ethical principles and norms to the decision-making process in agriculture.

* Emeritus Professor of Philosophy at the University of Costa Rica

P.B. Thompson (ed.) *The Ethics of Intensification*,

Moreover, there is often a vague perception of the existence of problems without a clear identification and characterization of them, usually because the values underlying the selection of problems and the proposal of solutions remain implicit. The shallow approach referred to above is often complemented with the idea that the arena for decisive action is not ethics, but politics. Ethics is thus seen as limited to the private sphere, whereas politics is conceived of as the realm of public and social issues. No wonder, then, that ethics is considered interesting but powerless and politics powerful but unethical. Likewise, politics is conceived as the realm of the possible whereas ethics is often considered a paralyzing request for the impossible.

None the less, let us try to place ethical problems raised by agriculture intensification within the larger context of socio-economic change and, therefore, in the public arena where plans and projects are discussed. Such is precisely the task of development ethics, which may be seen as a critical examination of the values behind the selection of problems and the proposal of plans for social change. For more than a half century governments and international institutions have promoted development as the means to overcome the overall condition of destitution that afflicts millions in vast regions of the world. Underdevelopment is accordingly seen as a situation that peoples and governments want to leave behind. Albeit development is sometimes seen as a permanent condition enjoyed by some countries, we assume here that it is a continuous and open-ended process of change. Underdevelopment is usually considered a situation of stagnation and deprivation that affects society unless a decisive effort is made to overcome it. Hence the need for government-sponsored development plans and planning agencies.

Although in the short run a temporary improvement of social indicators may take place when the prices of exports rise, the usual approach to development disregards sudden changes and focuses on long-term strategies. It is in this perspective that science and technology are considered necessary conditions for change because of their impact in the production of goods and services. So, the first step envisioned in approaches that otherwise differ greatly among themselves is the transformation of agriculture through the application of science and technology in order to obtain greater productivity with the corresponding increase in labor available for industry, which is supposed to be the real engine for economic growth.

Traditional agricultural societies, consequently, usually suffer the impact of change before other segments of the population and in a more lasting way. When social change is ethically examined, so that the good and bad aspects thereof become evident and open to discussion, one should bear in mind that development often begins with disruption in societies which place great value in permanence and stability. Change is deepest where it is least appreciated.

Thus, the initial increase in productivity in development processes is usually brought about by agriculture intensification, and therefore the assessment of the consequences of such intensification for people and the environment are very important in development ethics. Who benefits and how, as well as who suffers in the process, seem to be the first questions to ask. Answers will be complex because benefits and harms are widely distributed and because there seems to be no easy way to avoid harm to some while benefiting others. Development plans find legitimacy in the improvement of conditions they are supposed to bring about. A quick glance at the starting point and the results of the process will show some interesting asymmetries.

There seems to be a consensus in the description of underdevelopment in terms of lack of satisfaction of basic needs like food and clean water, clothing and housing adequate for the climate, education, health care, job opportunities, rule of law and good governance. Developing nations, by whatever names they are called, share similar problems of deprivation although different degrees may be noticed. The nations at the bottom of the list put forth by the United Nations Development Programme (UNDP) in its yearly Report have in common an overall condition of destitution, often reinforced in a vicious circle by civil strife, corruption and ecological degradation. In many of such countries there still is a large segment of the population devoted to subsistence agriculture. This kind of traditional society associated with subsistence agriculture is often the subject of romantic idealizations, but in the context of economic theories it is usually considered either the stumbling block in development processes or just the initial stage in a process of change. Coming to grips with such contradiction in perception and valuation is one of the points we want to explore in what follows.

If the starting point of the process is well known, on the contrary the goal aimed at is hard to describe. The reason is that it is more difficult to agree on what a developed country looks like. First, there are several models of development and the paths chosen by different societies do not coincide. The United States and Sweden, for instance, are both developed countries according to indicators like per capita income or life expectancy but there are obvious differences between them. Taxation in Sweden is much higher than in the United States but so is the range of social services. If asked what country they look up as the model to follow, people would answer in different ways. While there may be United States citizens who long for the degree of security enjoyed by the Swedes, some Swedish nationals may want to reduce their taxation burden to United States levels.

Second, it can be said without contradiction that there are many degrees of underdevelopment but no real development in today's world. If we hold that development is an ideal state beyond what social and economic indicators measure, then we can maintain without contradiction that no nation on Earth has achieved such a condition. There is always room for improvement. This is particularly true when nations with a high HDI are seen in the global context and the impact of their high levels of consumption in the world at large is considered. The case could be argued that the majority of countries may be paying a high price for the development of the few.

Third, development projects create new problems. It is well known that homelessness and clinical depression, for example, are more likely to be found in countries in advanced stages of change than in societies devoted to subsistence agriculture, where family ties are stronger.[1]

[1] For the increase in clinical depression in highly industrialized countries, see Lane (1998).

Finally, if the unit of analysis is not the nation-state but regions, we find a mixture of varying degrees of regional development and underdevelopment within most countries, especially those in the middle of the United Nations Development Programme ranking. In many developing areas and in almost every industrialized country social indicators are not the same in all regions, and the range of variation is sometimes very wide. Two conclusions can be drawn here: there is development in underdevelopment and vice versa; development is a continuum with degrees.

In addition to the fuzziness of the distinction between development and underdevelopment, the very idea of development has experienced an evolution in the last decades of the twentieth century as a consequence of both a deeper theoretical understanding and the answers to practical problems encountered by governments and international agencies. Even though development is a continuum through stages, it has not shown the same direction throughout all these years. In addition to degrees, it also has different possible outcomes. And even though underdevelopment may be seen as the limiting case of development, both notions are not symmetrical when seen in historical perspective. Whereas underdevelopment refers to similar facts and conditions ever since it began to be used, there has been an evolution of the notion of development. The major steps in this process can be characterised as follows:

1. Increase in per capita income as a result of economic growth made possible by industrialization (Rostow 1960)
2. Economic growth with social equity (Amin 1976)
3. Satisfaction of basic needs, in such a way that development becomes possible even with low incomes (Streeten 1981)
4. Sustainable or long-term development (Brundtland Report of the World Commission on Environment and Development 1987)
5. Freedom from misery and for the flourishing of human capabilities (Sen 1999)
6. Human security as measured, for example, in the index used by the Global Environmental Change and Human Security Project, also known as GECHS, presently based at the University of Victoria in Canada

It is interesting to note that a similar and converging evolution can be seen in the notion of culture, especially when translated into policies promoted by governments and organizations. From the concept of culture as a collection of objects and performances, the notion evolved to stress personal security within a group with shared values and customs, as explained for example in Jean Ladrière's famous essay (1977) *Les enjeux de la rationalité*.

Since agriculture intensification in itself is some kind of growth, we may consider it within the framework of development models based either on economic growth or on the further distancing from such a notion. For those in favor of the model of development as economic growth, also known as trickle-down model, market forces suffice to distribute benefits among all social groups without compromising individual freedom. Once derided as a simplification of the notion of development, this model has been resurrected today by the three basic theses of what is sometimes known as neo-liberalism and other times as the Washington consensus[2]: privatization, fiscal austerity and market deregulation. Critics of the model point out that without some sort of corrective measures, usually by governments, economic growth would tend to increase the gap between rich and poor. Instead of growth alone, they contend that we need growth with equity. Freedom becomes an empty word for those who define themselves in negative terms, as the victims of privation and oppression.

But equity may not be enough, say those who favor the notion of development as satisfaction of basic needs. They point out that growth by itself should not be a goal, for at least two reasons. First, basic needs could be satisfied with available goods and services if only these are better distributed, not only in the sense of reaching more people but also as a consequence of a change in priorities. Second, unlimited growth is ecologically unsustainable.

Both growth-with-equity and satisfaction of basic needs models include some degree of government intervention. These is why both face today the challenge of the reduction of state agencies and of governments.

The same can be said of the typical Latin American approach to foster development, known as Sábato's Triangle. Proposed in 1972 by the late Argentinian engineer Jorge A. Sábato as a means to connect science, technology and development by forging a strong bond between the public sector, the productive apparatus and the scientific-technological institutions (its three angles), the Sábato's Triangle is widely used by government agencies in Latin America as a strategy to establish the necessary connections for social change. Today, however, it faces the problem of having one of the angles – that of the government – severely reduced as a consequence of the triple policy of privatization, fiscal austerity and market deregulation.

In addition, there is a deeper problem: in many developing nations neither enlarged nor reduced governments have succeeded in creating the conditions for the satisfaction of basic needs of the majority of the population. Corruption is as much a problem for development as greedy market forces. Moreover, healthy markets need strong judicial systems that are usually absent in many nations.[3] So, instead of efficient private initiatives versus corrupt governments, many nations face both corruption and ineffectiveness at all levels. No wonder, then, that corruption and good governance have entered the discussion on development.[4] Organizations devoted to propose alternative models of development, like the International Development Ethics Association (IDEA), have devoted recent conferences to these topics as for example the International Conference held in the agricultural school El Zamorano, Honduras, in 2002.

A related challenge for philosophy coming out of discussions on development is to assess technological innovations from an ethical perspective. Since technology is today the realm of human action where change seems to be most prevalent and fast,

[2]On the Washington Consensus see, for example, Stiglitz (2002), especially p 67.

[3]Hence the title of the World Development Report 2002, *Building Institutions for Markets*.

[4]See Rapley (2002), ch. 6.

the consequences of new technologies for the well being of humans and nature pose urgent questions to ethics. In this context, let us go back to the transition from traditional communities to high consumption societies through the application of science and technology first to agriculture and then to manufacture. For many of us, traditional agricultural communities with their strong attachment to land and family was the beginning, the starting point in life and the reference by which personal security was measured. In just a word, it was home. Easy to romanticize, that kind of life was based on strong family ties which gave most people, otherwise mostly poor, a strong feeling of belonging and a whole set of values, among them frugality, solidarity, and appreciation for hard work together with the simple enjoyment of life.

Two conflicting attitudes prevail in more contemporary views of traditional agricultural societies. The first one is romantic. According to it, such a condition was the epitome of human life seen as a positive and personally fulfilling experience. The landscape was home for countless millions of human beings for several millenia, and some of the six-plus billions of human beings on Earth still remain close to rural and sometimes untamed landscapes. In developed nations the memory of such a stage in history has long faded from personal recollection in most people but is still present in other manifestations of something like a collective memory. In developing countries at least the oldest generation may still remember the farms of their childhood, and when recurrent economic crises nowadays hit such nations with great severity there may remain somewhere a small plot to plant some corn, beans and vegetables, and perhaps even raise some chickens and pigs. During the Lost Decade of the 1980s, when the external debt crisis inflicted on people in the then so-called Third World the double curse of inflation and unemployment, grain production in some Latin American countries actually increased. People were going back to the land, to the primeval relation between soil, plant and human being. Towards the end of Arthur Miller's famous play *Death of a Salesman* Willy Loman looks for a spot between buildings to plant a garden, as if trying to revert a life-time history of persistent personal failure amid the cruelties of the market forces, and find redemption in the relation with nature. That no nature is left to him to go back to is a big part of the tragedy.

Easy to idealize and hard to forget once lost, agricultural societies are usually associated with small villages instead of large cities, friendly neighbors instead of impersonal crowds, helpful relatives instead of preying strangers. Many of the critics of development long to go back to a world lost to industrialization, agribusiness and population explosion.[5] But subsistence agriculture, once an option when the population was counted in millions, seems inadequate to feed today's billions. Furthermore, our idyllic picture of traditional societies tend to be the effect of a selective memory, forgetful of the least appealing aspects of that life.

The second attitude towards traditional agricultural societies is just the opposite of the previous one, and tends to consider them merely as a stepping stone towards

[5]Such seems to be the message in Sachs' *The Development Dictionary, A Guide to Knowledge as Power* (1992).

more advanced phases of development. In *The Communist Manifesto* (1848) Marx and Engels clearly state that the only revolutionary social class is the proletariat created by the Industrial Revolution, not the community of farmers and peasants or the serfs of feudal relations. Progress is thus in the hands of people that do not work with nature, but with machines. In the opposite corner, W. W. Rostow in his 1960 epoch-making book *The Stages of Economic Growth, A Non-Communist Manifesto* seems sometimes to have a similar attitude towards agricultural communities as he lists, among the necessary conditions for the take-off, value changes in traditional societies and in particular the acceptance of modern science and its use to increase productivity in order to free hands for industry.

So, in both Marx and Rostow – although in different degrees- traditional society is valued only or mostly as a link in the long chain of progress. The horrors of collectivization in Stalin's Soviet Union had a theoretical underpinning. Let us hope that no one sees agricultural societies in such a way today, when one of the main ethical challenges is to assess agriculture and agricultural communities for what they are in a larger picture. Food production is all-important, but different actors in the process have different participation in the process, and consequently should have a different degree of participation in the distribution of benefits. A proper assessment would thus avoid the distortion through manipulation of values we see today. Without farmers there is indeed no food – as the oft-repeated slogan proclaims – and without food there is no life. But rich landowners and agribusinesses use this slogan in order to secure privileges that frequently do not translate into benefits for their ill-paid employees. That this slogan is seldom used by the weakest link in the chain, the landless peasants that sell their labor to landowners, should be no surprise: they lack the political clout to demand a larger share in benefits, and the usual way so far to improve their conditions is government intervention, today greatly reduced as a consequence of the idea that a good government is a small and powerless one.

In addition to a proper assessment of the place of agriculture in human life, ethics thus faces the challenge of assessing the impact of technology in agriculture. Agriculture intensification is usually possible through the application of technology, but the use of technology is not always a good thing. Applied to agriculture, technology may bring about a betterment of conditions for most people but it often leads only to larger profits for some on the excuse of the benefit to all. A worsening of overall social and ecological conditions may also follow. Although technology is usually associated with the improvement of conditions in which humans live, it has become in many occasions an instrument of oppression and destruction.

It is easy to find examples of technologies fundamentally flawed and inappropriate for the purposes they were supposed to achieve. Their imposition on society was in many cases the result of political decisions lacking in technical and scientific grounds. As is well known, Lysenko's agricultural methods backed by Stalin resulted in widespread famine and in the elimination of dissenters. Mao's attempt at the industrialization of China during the period known as the Great Leap Forward (1958–1963) was not only a great failure and cause of starvation for millions, but also meant the destruction of the few forests remaining in China at that time.

In many Latin American countries the introduction of short varieties of coffee grown without protective shade led to the elimination of the traditional family plot where many species of trees, vines and bean plants grew together to form a complex and self-sustained forest. When the traditional plot was replaced by modern plantations, coffee productivity per unit of land increased greatly but biodiversity and food security for many poor farmers were lost. Today, in the aftermath of the plunge of coffee prices and in search for the high value of exotic brands, in many regions the old high-quality coffee plot with its array of chayote squash vines, bean stalks and fruit trees is making a strong and most welcome comeback.

Although wrong technologies can be corrected or abandoned as soon as their inadequacies are clearly known, some societies and political regimes seem more reluctant to change course than others. The rigidity of totalitarian regimes is often mentioned in this respect, but this is not the only source of opposition to change. When market forces prevail, without interference by either governments or groups of the civil society, changes in production and consumption patterns become more difficult.

The word "evil" applied to "technology" comes to mind when we see artifacts and processes used to commit genocide or to deprive people of freedom and opportunities. As the legal frame for the protection of industrial and intellectual property, the whole system of patents, rights and franchises internationally enforced today as seen by many in developing nations is approaching a degree of control without precedent or justification in history. In the past, farmers often were able to provide for themselves most of the inputs of production, from seeds to fertilizers. When they needed to buy these things they could choose between competitors. But if every one of these inputs falls in the hands of monopolistic corporations, very basic aspects of human freedom and security are lost in the process. It is useful to remember that today's advanced nations profited by the free flow of information, discoveries, inventions and experts that was the norm in the past. In the frequent conflicts between those nations any possibility of catching up with technological innovations was considered morally justified. The situation has changed dramatically for developing nations, and one of the tasks of development ethics is to analyze the asymmetrical relations between rich and poor countries.

The ethical analysis of technology transfer must be seen in connection with the more general problem of the ethics of technology in itself.

Many defend the idea that technology is ethically neutral in the sense that ethics becomes relevant only when the user makes decisions concerning the application of objects and processes. The issue of responsibility would thus arise only when technological objects and processes are available to potential users, not before. The standard objection against this contention goes like this: since the possible uses and abuses of technology can be taken into consideration already at the design stage, the question of aims and purposes, of good and evil, arises even before the product comes into being. This applies in particular to highly specific technologies. A knife can be used to carve a turkey or to commit murder but atomic bombs have very few possible uses. So called "Terminator" seeds may be very efficient to increase crop yields, but they are designed to strengthen the dependence of farmers on an all-powerful corporation whose gain is legally protected through the patent system.

The impact of any new technology on the environment is another question that should be answered already at the design stage. In agriculture this applies not only to technological objects and processes used in the production of food, but also to the expansion of such activities into new lands.

Similar to artifacts, development plans are designed for specific purposes in accordance with certain values, though these are often implicit. Who makes decisions, and on which grounds are always ethical issues of great importance. As pointed out by Denis Goulet,[6] in any decision concerning development three aspects are ethically relevant: the selection of problems to be solved, the implementation of the plan and the way the solution is worked out. These aspects of a single process reveal the values held by decision makers, and development ethics aims at making those values explicit as a way to avoid that the emphasis on the technical character of planning become the excuse to avoid consultation with those who suffer the consequences.

In working out such concern, development ethics may follow two paths. The first one stresses the role of the state as the proper instrument to achieve a just society. Beyond commutative justice, where personal differences are considered irrelevant in transactions between individuals who are thus supposed to be equal partners, distributive justice aims at equality among people in unequal conditions. The state is considered to be the proper instrument in achieving distributive justice by looking after the interests of the poor and weak as opposed to the gains of the strong and powerful.

But many people see little relation between public policies and distributive justice in modern states. They point out that politics is often conceived of as the art of grabbing and keeping power. Distributive justice, in fact any kind of justice, seems to be outside of the ordinary concerns of rulers and states. No wonder, then, that most people in the world do not relate justice with the actions of the ruling classes and instead have lots of reasons to see them as insensitive to justice.

Since development plans imposed on the masses in the name of a particular ideology, or based on purely technocratic criteria often lead to unrewarded suffering for most people, development ethics is actually called to fill a vacuum. Its role is to provide a critique of unexamined ends and means implicit in plans and decisions, to become a voice for the victims of development projects and a calling for the accountability of those who place themselves above the ordinary citizens.

It is here where an analysis of the definition of development will help us to understand the role of development ethics.

In its social and economic sense and in the most general terms development is prima facie an increase in income or consumption per capita plus some kind of social change. The first aspect, usually referred to as economic growth, is easy to measure but can be sought for purposes other than the improvement of conditions of the population. The welfare of millions of people, for example, have often been sacrificed to the military objectives of their countries. The history of Europe, in particular, is full of such cases.

---

[6]Goulet explains his method, for example, in *Development Ethics, A Guide to Theory and Practice* (1995), ch. 1.

Social change is more difficult both to define and to measure, and in fact has been reinterpreted as the idea of development evolved as already shown. However, no matter how development is defined, some degree of improvement of human conditions is implicitly or explicitly included in interpretations and reinterpretations of the term. In models based on growth the improvement of conditions is taken for granted as a result of rising income. In models based on satisfaction of basic needs, freedom or security, better quality of life is placed in the forefront from the start.

Similarly, the special role of technology in both economic growth and social change was obvious from the beginning and is to be found in different models.

In connection with the first aspect, in W.W. Rostow's *The Stages of Economic Growth* (1960), one of the conditions for economic take-off is what he calls "the fruits of Newtonian science," i.e., scientific theories, technological tools and know-how. It is technology applied to agriculture what makes labor more productive, thereby freeing large numbers of hands for work in industry, where technology makes possible large-scale manufacturing.

At the same time technology plays an all-important role in the second aspect of the notion of development, social change. It is well known that new technologies eliminate many old jobs, but they also create new ones, as well as new businesses, institutions, possibilities and even life styles. Sociologist David Freeman in *Technology and Society* (1974) talks of four successive phases of the impact of technology in society. First, new artifacts simplify daily tasks and eliminate old problems. To mention an obvious example, one of the fortunate consequences of using a word processor rather than a typewriter is that correcting mistakes in carbon copies is now something of the past. In a second stage, job qualifications change even for traditional jobs. Secretaries are now expected to know how to use computers and handle electronic mail. Third, the way authority and prestige are distributed among social groups also changes. Since people trained in new technologies are in high demand and therefore make more money than those working in older technologies, they likewise enjoy greater social status. The authority and prestige enjoyed by primary school teachers, clergy and policemen in traditional societies is no longer the same in industrialized countries.

In the final stage many values that guide social behavior also change. For example, values of subsistence-agriculture societies and those in industrial nations are not the same. Perhaps the most striking change is the one that has taken place in consumption. In traditional societies, where productivity is low, frugality is a virtue. In mass-consumption societies, where high levels of employment depend on consumer confidence, frugality is a problem.

To Freeman's four phases of change we can add the idea that even the valuation of change itself is subject to modification. In traditional societies any change is viewed with suspicion even when its advantages are clear. In modern societies any change is highly valued even when its advantages are dubious; this may explain why some religious groups freeze technological change at a given stage. Since technology influences morality by changes in valuation, there is ample room for ethical analysis of social change brought about by technology. It is an open question which changes are desirable and which are not. The fact that a new technology makes possible greater productivity and new opportunities for individuals does not mean that it is better in all respects. It may increase the gap between the rich and the poor, and therefore contribute to social conflict, or be more destructive of nature. Higher crop yields is usually associated with monoculture, whereas many traditional practices, with their typical combination of different species, were safer both for human beings and for the environment.

This is one of the reasons why old technologies and techniques should not be forgotten, so that their surviving practitioners should be encouraged to teach others. Once a practical solution for a problem has been found – no matter how primitive it may seem to us today – there is no way to tell when and how we may need it again. It is well known that in times of crisis knowledge of old ways of doing things may represent the difference between life and death for whole groups. In addition, each technology requires some conditions for its functioning and more advanced technologies usually require more specific inputs. When such conditions are not present the possibility of using alternative technologies is crucial for survival, and older ways of doing things may be the best solution.

The relation between science and technology has been and is still today the subject of a controversy which is relevant for development, although perhaps in an indirect way. It is clear that many technologies precede the corresponding sciences, and this is true not only of metals, glass and textiles, but also of more recent inventions. James Watt's steam engine was prior to thermodynamics and the first successful airplane was built by two bicycle mechanics, Orville and Wilbur Wright. Nevertheless, there is a strong link between modern science and contemporary technology and a good mastery of natural science and mathematics is considered a necessary condition for the advancement of technology. In fact, a strong public commitment to scientific and technological research is considered a necessary condition for the overcoming of underdevelopment. But since underdevelopment is a social condition, social science is required in the search for its explanation. Unfortunately, it is typical of development plans in developing countries to leave little or no room for the promotion of social sciences.

Confirmation of the need for both natural and social sciences can also be found in the history of theories on this subject. Development economics traces its roots to authors like Francis Bacon (1561–1626) and Gottfried Leibniz (1646–1716), who sought the type of knowledge of nature that could alleviate human misery, but also to David Hume (1711–1776) and Adam Smith (1723–1790), who discussed the difference between rich and poor countries and whether it was morally desirable to bridge the gap. Both quests ended in affirmative answers: Bacon and Leibniz provided cognitive approaches aimed at alleviating the miseries of a brutish and short life lived in constant conflict; Hume and Smith considered morally desirable to bridge the gap between rich and poor countries. Contemporary approaches to development inherit trends from both natural and moral philosophy.

After the upheaval of the Industrial Revolution in Great Britain in the late eighteenth and early nineteenth centuries other countries experienced similar and equally profound economic and social changes. It is in the twentieth century when countries with different political regimes formulated and implemented large economic plans, usually for political reasons: the Four-Year Plans in Germany and the Five-Year Plans in the Soviet Union in the 1930s, as well as F.D. Roosevelt's famous 100 Days with a large number of measures destined to overcome the Great Depression. It was in the 1950s when the obvious differences between two kinds of countries entered the political arena on an international scale and when explicit plans devised to make things change became common practice among governments and international institutions.

Soon it was clear that development plans were not helping large numbers of people forced to change in the process. Change without benefit for those affected is not the same as development, just as development that cannot be sustained is not the same as progress. A critical examination of the ends and means of development, the values implicit in plans and the real beneficiaries of change, was undertaken by several authors. As already pointed out, in many publications since 1965 Denis Goulet has worked out a method to do so, which examines the choice of problems and solutions in the light of values implicit and explicit. Development ethics as conceived by authors like Goulet and David A. Crocker aims at proposing alternative models to development. The respect for cultural values – and, therefore, for plurality and difference – is essential in such models, with the expectation that in this way social changes brought about by development will have deeper roots and be, consequently, more sustainable in the sense that the next generation will find at least the same natural, man-made and human capital than the previous one.

The preceding considerations can be posed in more general terms: how can development plans be morally justified?

Assuming a universal right to a good quality of life based on human freedom, there is a general and simple answer. Development plans and projects begin to include a moral justification when human beings are no longer considered mere objects and become subjects in the decision-making process. Because of their instrumentality in connection with development, science and technology are also affected by the process of becoming human actors instead of passive recipients. To begin with, scientific theories must be relevant in the solution of the problems of the dispossessed. In particular, an approach in economics which fails to appreciate the importance of problems like unemployment and asymmetrical relations in trade is therefore considered morally defective. Instead of theories insensitive to the special needs of developing regions of the world, a good economics of development is required. Second, the next step is to create an economics *for* development. When non-economists concerned with the unmitigated evil of poverty and unemployment look for answers in theories of economics what they find often seems irrelevant. Other social sciences, especially psychology and anthropology, are also called to contribute to development. Obviously, the resources of natural science should be harnessed in the effort to increase productivity and reduce poverty.

A further step in the move from passive recipients to active subjects concerns the formulation of development plans and projects. Local knowledge, techniques and technologies are often more appropriate than imported ones. This is especially true in regions where technology is usually seen in connection with foreign domination. As aptly pointed out by Jane Robinett (1994), Latin American fiction – unlike the literature of other regions of the globe- usually associates technology with foreign exploitation and deceit ("this rough magic," in the title of Robinett's book, taken from Shakespeare) and places a high value in local knowledge. This point is beautifully made in Jorge Amado's novel *Gabriela, Clove and Cinnamon*, where Gabriela's wonderful ability to prepare delicious dishes with available foodstuff is compared to the dependence on imported items shown by a French chef brought into the town of Ilheus. Towards the end of Gabriel García Márquez's novel *One Hundred Years of Solitude*, the rusting chassis of a car in a muddy field without vegetation is all that remains of what once was a lush forest, after the foreign banana company has destroyed both town and jungle. At the beginning of the novel the wandering gypsies introduce ice and magnets to Macondo, and these quickly become objects of appreciation because of their magic qualities in the eyes of the locals. Magnets, for example, are seen as capable of awakening the hidden life of inanimate objects. But the installation of railroad tracks makes possible the arrival of the predatory banana company, and with it the doom of people and nature.

When respect for culture and its values is genuine, development plans are not imposed from above but become the outcome of taking into consideration aspirations and desires of the actors involved. Although values deeply ingrained in cultures may be inimical to many things associated with development, we assume here that the struggle to alleviate misery and pain is not restricted to just a few human groups. Some societies may be more willing to experiment with change than others, and different attitudes toward development may create conflicts among groups. But cultures are neither frozen in time nor free of contradictions. Thus, paying attention to the victims of values held by the majority but denounced by many in the community may be the way for an overall improvement of conditions.

Nevertheless, to move from passive recipients to actors in development is not a sufficient condition for success. It is obvious that projects and plans rooted in local values and resulting from negotiation among groups may be morally or technically defective. There may be consensus on the implementation of non-sustainable projects; individual greed could be one of the ingredients of public agreements, and corruption is sometimes the glue that holds together social arrangements. Thus, democratic participation by itself does not guarantee that a decision is morally right and technologically appropriate. In avoiding situations where the public good is compromised some help may be found in the interplay between two kinds of actors in development, called *insiders* and *outsiders* by David A. Crocker (1991). Their role is different but complementary: insiders should be able to incorporate local values into the process; outsiders may distance themselves from them when judging about the rights and wrongs of decisions made by insiders. Thus local experiences, relevant but limited, may be complemented with outside expertise, less specific but wider in scope. The same person may begin as an outsider and later become an insider, or may be an insider in relation to a development project and an outsider relative to another one. Both insiders and outsiders are needed in development projects, although for a fruitful encounter to occur both groups must share basic values related to the improvement of human conditions.

The objection has been raised that in a world where 24-hour television gives us instant coverage of events taken place anywhere, we all know what is going on everywhere. But even in a globalized world profound cultural differences are still with us, so that the distinction between insiders and outsiders remains valid. Communication between both groups helps to make explicit the three aspects of development projects: the choice of problems, the proposal of solutions and the values behind both.

By providing some ideas on how to do it, development ethics facilitates a dialogue among different groups willing to share factual experiences and moral insights in the struggle for human survival. Because of its unique role in such a survival, the answers given to questions arising from agricultural practices are of vital importance to all of us.

Chapter 7
Comments on Luis Camacho, "Agriculture Intensification from the Perspective of Development Ethics"

Stephen L. Esquith*

Is there a distinctive "ethics for development," or is development just one of a number of pressing subjects to which ethical theories and methods of analysis ought to be applied? This is the philosophical question that Luis Camacho has raised. It is one that forces philosophers to confront some of the ways in which ethics borders on metaphysics and epistemology. It also has important implications for practitioners in the field of development, including agricultural development.

Camacho's question has parallels in other areas of practical philosophy. Not long ago, moral philosophers were asking similar questions about medicine. That is, is there a *sui generis* medical ethics whose particular subject matter requires a special set of ethical tools and whose application requires an ethical point of view different from the dispassionate and removed point of view of the conventional ethical theorist? For example, are there features of patients who suffer from particular diseases and illnesses, say Alzheimer's disease, that make the standard distinctions that philosophers have made between conscious and unconscious beings difficult to apply in this special case? Or, does a patient in a persistent vegetative state present special, even insurmountable problems for an ethical theory premised on standard views of what it means to be a person with certain rights and duties? Who is to decide for patients in these situations? Questions such as these have spawned a robust academic discipline of medical ethics, with its own academic journals, professional associations, and clinical and casuistic methods of ethical reasoning. Is this what lies ahead for moral philosophers concerned about the ethical problems that arise in the process of development?

Luis Camacho argues that a distinctive "ethics for development" is already emerging, and it consists of mid-level principles driven in large part by radical changes in technology. I suggest that there is another way of interpreting the emergence of the field of development ethics that is consistent with Camacho's principles *and* the more general ethical theories to which Paul Thompson has called our attention. Practical ethics, whether it is in law, medicine, or agriculture, must rest upon plausible metaphysical commitments and reasonable epistemological

* Dean of the Residential College in Arts and Humanities at Michigan State University

P.B. Thompson (ed.) *The Ethics of Intensification*,

presuppositions. To be persuasive, these commitments and assumptions must be part of a more general ethical theory, but the linkages between mid-level principles like Camacho's and the general ethical theories, whether they are consequentialist, deontological or virtue-based, need not be tight. That is, more than one general ethical theory may ground the same set of mid-level practical principles. This makes practical compromise more likely without sacrificing integrity and philosophical coherence.[1]

There is another sense in which Camacho's notion of an "ethics for development" raises questions about the adequacy and relevance of traditional ethical theory. He asks us to think not about the abstract values of justice, equality, and liberty which traditional ethical theories have concentrated upon, but rather something called an ethical "way of life." An "ethics for development," according to Camacho, requires that we see how these values fit together for particular people, that is, how they define their way of life. This notion of an ethical way of life is not foreign to traditional ethical theory. In fact, it is reminiscent of Isaiah Berlin's formulation of the true subject matter of ethics.

> Ethical thought consists of the systematic examination of the relations of human beings to each other, the conceptions, interests and ideals from which human ways of treating one another spring, and the systems of value on which such ends of life are based. These beliefs about how life should be lived, what men and women should be and do, are objects of moral inquiry; and when applied to groups and nations, and, indeed, mankind as a whole, are called political philosophy, which is but ethics applied to society.
>
> (Berlin 1992)

In a world in which village life and industrial agricultural production are linked across seemingly centuries of difference, the ethics of our ways of life must also be connected. We are, as David Held has stressed, not citizens of discrete and insular societies, but members of "overlapping communities of fate" (Held 2003). In this context, for ethics to be a "systematic examination of the relations of human beings to each other," it must be an "ethics for development."

In this brief comment I will be trying to accomplish three things. First, I will argue that one reason why we need a distinctive "ethics for development" is that globalization has created a new set of constraints on our ways of life. We live in a world that is simultaneously integrated and fragmented in unprecedented ways. This does not make traditional ethical theories of justice, liberty, and equality obsolete, but it does require that we enlist these theories in very different ways in order to ground the mid-level principles Camacho would have us consider. Second, I will try to make more explicit these three principles that I believe Camacho commends to us. Third, I will extend Camacho's argument beyond the subject of the proper ethical conduct of moral agents to the practical issue of collective responsibility. I believe that an "ethics for

[1] John Rawls has discussed this matter extensively and coined the term "overlapping consensus" to describe the way that he believes different "comprehensive moral theories" may overlap in their support for political theories that remain agnostic on controversial metaphysical matters. See, for example, Rawls (1993). On the subject of compromise and integrity, see Martin Benjamin (1990).

development" must address the subject of organizational and institutional responsibility. Standard ethical theories have not moved much beyond individualistic notions of weak positive duties of beneficence and stronger negative duties of justice. There is a range of responsibility between these two poles along within which the responsibilities of groups, organizations, and institutions lie.

The Context of Globalization

Throughout 2003 *The New York Times* ran a series of lead editorials entitled *Harvesting Poverty* that described the relationship between United States, European, and Japanese food policy and the conditions of hunger and poverty in other parts of the world. The series stressed the negative effects of U.S. farm subsidies and import quotas on the United States economy as well as on producers and consumers in developing countries.[2] The editorial on December 30, "The Unkept Promise," began with the following gloss on globalization.

> There is a deceiving sense of timelessness to the stillness of rural life. The jungles of Mindanao offer few clues as to whether it's the early 20th century, or the early 21st. Nor do the highlands of Guatemala, the Mekong Delta in Vietnam or the cotton-rich plains of the Sahel in West Africa. But these disparate regions are very much of the present, stitched into the quilt of global commerce. World trade links us to them, as surely as it links London, Tokyo and New York.

The "quilt of global commerce" does indeed patch us together in complex ways. Some parts of everyday life in developing countries remain rooted in an earlier epoch at the same time that the citizens of those countries travel by Internet and satellite connections to far corners of the globe. Knowledge and resources can be simultaneously geographically far away and electronically ready at hand. This is what James N. Rosenau has aptly described as "distant proximities." He has argued that the pattern of globalization connecting societies at different points of development is both integrated along certain vectors and fragmented along others so that what is distant geographically may be proximate in other ways, and vice versa (Rosenau 2003).

As science and technology have been deployed to solve problems of hunger along these complex global vectors, they have also heightened the tensions between subsistence agricultural ways of life and large-scale agricultural production. This is perhaps most noticeable in the controversies surrounding the introduction and regulation of genetically modified organisms. In addition to concrete questions about legal liability, intellectual property rights, and risks to the environment and future

---

[2] "*The Times's* editorial page is focusing on the damaging impact that American, European and Japanese agricultural subsidies and trade barriers have on farmers in developing nations. The project is being led by editorial writer Andrés Martinez, who is travelling to Asia, Africa, Europe and South America to research the issue." <http://www.nytimes.com/ref/opinion/harvesting-poverty.html?pagewanted=all>

generations, there is also the matter of how new technologies such as these can transform ways of life. In other words, there is both integration in an economic sense and dangerous fragmentation socially and politically, and this is what requires that we think about the possibility of a new *sui generis* "ethics for development," not merely the application of existing standard ethical theories to problems in development. We live in a world in which our ways of life overlap, but our ability to forge a consensus about ethical issues has lagged behind.

Technology and Practice-Driven Theory

An "ethics for development" does not take as its primary object of investigation the contested notion of development, at least not directly. 'Development', Luis Camacho stresses, is too ambiguous and elusive a concept. If ethics is to advance development in a practical way, then it must recognize this. Homelessness and clinical depression, for example, are symptomatic of 'developed' countries, and the very idea of 'developed' as opposed to developing countries is obscure. (Is the United States or is Canada a developed country?) Instead of trying to define development ethics based on the correct meaning of development, he prefers to take a closer look at the historical relationship between technology and our ethical ways of life as they have developed unevenly and ambiguously over time. This is why I have labeled Camacho's "ethic for development" a practice-driven theory of development ethics.

Camacho suggests that agricultural technologies used for greater production and efficiency are not neutral tools. Like all technologies, they have unintended consequences (e.g., the harm to biodiversity that may accompany greater productivity). This is "bad" technology, he tells us, but it is not the worst. There are also "evil" technologies that are intentionally designed to do harm. In an aside, he compares the terminator seed to the torture chamber: neither is a neutral technology that is potentially harmful when it is used for the wrong purposes. They are harmful technologies precisely when they are used the way they were designed to be used, that is, for enslavement and exploitation. Evil technologies are also not only evil because of their intended harmful consequences, he argues, but just as importantly because those who may suffer the intended harmful consequences from them not surprisingly have not been part of the process of designing these technologies in the first place. Potential victims of new technologies should have a voice in the political process through which these technologies are designed and deployed.

New technologies also must be assessed in terms of their social and cultural effects, not just their effects on economic growth. Technologies that improve production also create new pressures on workers to master new techniques. These new techniques can simplify daily life at the same time that they can complicate job hierarchies in the workplace and re-arrange the distribution of social status. Even the small virtues of daily life (e.g., "frugality") can be transformed and, of course, the cultural identities of particular groups can be threatened to the point of extinction.

Development ethics ought to take notice of harmful social and cultural consequences of new technologies that intensify the production process because they ultimately may undermine the very economic development new technologies seek to achieve. There is, in other words, a "bite back" effect. Designed to improve material life, they may so reorganize social and cultural life that material well-being is subsequently harmed because of social conflicts and cultural hatreds (Tenner 1996).

How can these two goals (i.e., avoidance of the harmful effects of bad and evil technologies, and attention to social and cultural impacts as well as economic effects) be achieved? Some cultural traditions ("local knowledge") must be protected, not for their own sake, but because without them development will not be agent-centered. That is, without their own cultural identities, groups that should be stakeholder participants in the process of designing technology, will not have the cultural resources to participate (Kymlicka 1995). Not all local knowledge and traditional values ought to be accepted in this process of technological design, but without broad cultural participation in the design and evaluation processes the new technologies will not be *both* more efficient *and* appropriate.

In order for members of developing societies to adopt and adapt new technologies that are more efficient and appropriate, and that are neither bad nor evil by design, there must be a dialogue between those David A. Crocker has called insiders and outsiders. This is more than a conversation over the most efficient or least disruptive technological innovations. Camacho argues that it is a "dialogue between cultures" – a new complex way of life – in which those with local knowledge are part of a conversation with those on the leading frontiers of the industrial world. To summarize his argument thus far,

1. Potential victims of new agricultural technologies that intensify production should have a voice in the design as well as the use of these technologies to avoid bad and evil consequences.
2. These harmful consequences can extend to social relations and cultural identity, not just economic well-being. Without the secure cultural identity that is put at risk by some new technologies, some groups may not be able to participate effectively in the technological design process.
3. For this dialogue between traditional and industrial cultures to result in both more efficient and appropriate technologies for agriculture intensification, there must be a method for balancing the contributions of insiders and outsiders. The rules of participation must be fair to all those who are affected and cognizant of the valuable contributions of local as well as expert knowledge.

These three conclusions can be restated as regulative principles of (1) inclusion, (2) identity, and (3) discourse,[3] and together, they form a Habermasian discourse

[3] The 2003 *Report of the FAO's Panel of Eminent Experts on Ethics in Food and Agriculture* argues that agricultural intensification should ensure access to all, avoid harmful ecological effects, involve all the relevant stakeholders in the design and implementation processes, and take the interests of future generations into account. <http://www.fao.org/DOCREP/005/Y8265E/y8265e00.htm>

ethic for development (Habermas 1998). The first principle holds that all those persons affected by technologies that intensify agricultural production have a right to be included in the design and implementation process. The second principle prohibits the colonization of the traditional life world of developing societies by technologies imported from industrialized societies because they corrupt moral community and virtuous character. The third principle is a principle of democratic governance or contractual fairness that limits the authority of professional and technical experts in the decision making process.

All three of these principles eventually may be supported by the general ethical theories that Paul Thompson sketches. There may be, in other words, good consequentialist, deontological, and virtue ethics reasons for subscribing to these principles of inclusion, identity, and discourse. However, this kind of theoretical support structure will not precede the process of negotiation and compromise that Camacho's discourse ethic authorizes. There is no pre-existing overlapping consensus, to use Rawls's language. Only gradually, as new groups are included, identities reformed, and expert knowledge demystified will there be a recognition of mutual dependence and an emergent sense of belonging to "overlapping communities of fate." This leads to one further question: what responsibilities are entailed by membership in these overlapping communities?

Responsibility

Luis Camacho suggests that new agricultural technologies are not merely the products of discoveries and advancements in the natural and social sciences. They also may have a reflexive relationship to science. In the past, he reminds us, new technologies sometimes have been the impetus for advancements in science. In the future, efficient and appropriate agricultural technologies may be able to kindle a new normative science of "economics for development" in which development is understood in a diverse and participatory way. This perspective of development ethics eventually may extend beyond a narrow economics *of* development that too often has been at odds with social and cultural well-being and with political democracy.

The idea is that technology, including technologies for agriculture intensification, have expressive-symbolic significance as well as the instrumental capacity for good, bad, and evil consequences. These technologies enable us to express new dimensions of ourselves, not merely to discover ideas, feelings, and thoughts that were there already but somehow unable 'to come out' (Pacey 2001). In other words, technologies can create new ideals of self, other, and society, just as they can discover new data and regularities that force us to revise our scientific theories of what is. The technologies do not miraculously do this independently of human activity, but they may organize and guide human activity in determinate ways that allow the creation and expression of individual and social identity. This expressive function of technology is most obvious in cases of artistic and musical technologies, but it is true for industrial, agricultural, and information technologies that (1) set a new tempo in the workplace, (2) refract the governing images through which we see ourselves and others, and (3) change the limits on who we believe is "one of us." New biotechnologies have changed our views of our kinship with other living creatures. New information technologies have changed what it means to make decisions from a distance as outsiders. It is not just the hum of the engine to which we become attuned. It is also the speed of the simulation and the capacity of the cyborg to converse with us that change how we express who we are as a people, and equally important, who we are not.

This last point about collective identity bears repeating, and it brings us back to the sense in which our potentially ethical ways of life may overlap. New technologies designed to intensify production also may express what "overlapping communities" hold in common that makes them who they are. These technologies can be ways of organizing work and ownership, not just conventional tools and machines. For example, intellectual property rights, just like rights to land, sea, and air can protect and clarify the meaning of common identity or they can erode and transform it. IPRs express who we wish to become just as much as they enable us to protect older conceptions of ourselves. What has happened, in fact, is that rights to conventional forms of property are now inseparable from intellectual property rights. The right to the code is as important as the right to the material on which the code is written, and it is not clear where one ends and the other starts. Technology is not just an engine of development. It enables those who use it to express themselves and their social, religious, and cultural values more or less effectively to one another, at the same time that it may deny this expressive power to others. In the case of agriculture intensification new biotechnologies and property rights in them can erode ethical ways of life, or they can be fashioned to affirm some of the overlapping communities of fate in which we find ourselves living.

This mixed result is unavoidable. Technology has the capacity to enable some to create, not just discover, their own identities, while blurring, if not erasing, the identities of others. Whether the conversation over the basic purposes of development will extend from the Sahel to Savannah, Georgia, and back again in no time, is unclear at this point. While the technology is available, the links and nodes we choose will be just that, a matter of choices that will structure a way of life in which local and global communities will no longer be measured geographically. We live within a complex network of power of our own making. Whether we can hold a "dialogue between cultures" that leads to ethical ways of life that are more inclusive, democratic, and respective of human agency within this complex network of power also remains to be seen. This is neither a matter of trying to pour old wine into new bottles nor even, as Luis Camacho suggests it might be, a matter of dressing up "traditional ethics in new clothes." It is a matter of expanding the categories of ethical theory so that they address the issue of responsibility in as nuanced a way as they address issues of inclusion, identity, and deliberative discourse.

We can begin by noting that consequentialism in an age of fragmentation *and* integrated globalization must be defined more broadly than earlier, mostly utilitarian ethical theorists have defined it. The deployment or withholding of new technologies will affect the fair value of human rights as it influences the distribution of material benefits and burdens. The ability to exercise rights, and therefore their meaning and value, is a consequence of basic institutions and practices just as much as the distribution of food, clothing, and shelter are. Similarly, the impact of technology on human character, including the capacity to act virtuously, is also part of the consequences of development technologies. Consequentialism, when it is defined narrowly in terms of expressed preferences or pleasures and pains, only masks these additional consequences for the practical value of rights and the flourishing of moral character.

Given this broader understanding of consequentialism, then, we can ask who is and who should be responsible for the good, bad, and evil consequences of development technology. As I have said, it is no longer possible to treat this question as if it applied to individual actors living in insular, self-sufficient societies. When consequences are defined more broadly and we recognize the integration and fragmentation of global networks, we are forced to ask how groups of people, institutions, organizations, and states themselves, not just individuals, ought to be held responsible and ought to take responsibility for the consequences of development technologies.

To close, let me suggest how we might begin to fill in this picture. The standard models for assessing responsibility are individualistic. From a narrow consequentialist perspective, we might say, those who caused the harm ought to pay the price. Similarly, from one deontological perspective, those who failed to meet their weak positive duties of beneficence or strong negative duties of justice ought to be held responsible.[4] These models strike me as inadequate for the ethical problems that arise in the context of development, including the context of agriculture intensification. For example, here we want to know what the responsibility of corporations who market certain products may be, if there is widespread harm but no clear line of causation connecting perpetrator and victim. Product liability in recent years has moved in this direction. Even where causation cannot be firmly established (i.e., it is not clear which manufacturer produced the particular product that harmed particular customers, but it is clear that some of the products harmed members of this class of consumers), some believe that liability still exists. Encouraged by successful class action product liability suits, there have been attempts to extend this notion of responsibility to human rights cases.[5]

---

[4] Some authors have suggested ways of conceptualizing group responsibility that is not reducible to the sum of the responsibilities of the individual members of the group, however they have tended to focus on trivial cases (e.g., groups of joggers) and have not discussed what burdens group responsibility for more serious matters carries with it. See Carol Rovane (1998).

[5] For a discussion of one aspect of collective responsibility arising out of the Nazi holocaust, see the class action suit brought against the German corporation by victims of slave labor, *Burger-Fischer v. Degussa*, 65 F. Supp.2d 248 and the subsequent compromise settlement brokered by the United States, *Agreement between the Government of the United States of America and the Government of the Federal Republic of Germany concerning the Foundation "Remembrance, Responsibility, and the Future,"* Berlin, Germany, July 17, 2000.

From a broader consequentialist view, however, I believe that we can ask a very different kind of question about organizational and institutional responsibility. If we think of ethical consequences in a way that encompasses utility, rights, and virtues, that is to say, an entire ethical way of life, then an exclusive fixation on individual perpetrators and victims can be avoided, and collective responsibility can be discussed without attributions of collective guilt. This is where the notion of complicity comes into play.[6] Without the organized support of many contributing members, an ethical way of life cannot be sustained. Conversely, without the complicity of many others, the erosion of an ethical way of life will not occur. Complicity may come in the form of negligence before the fact, but it may also come in the form of accepting and using the benefits of practices after the fact.

The claim that there can be accomplices before and after the fact to the abuses of power and failure to take due care in the process of agriculture intensification will be controversial. Are the consumers of products grown under hazardous conditions or with hazardous materials in developing countries really accomplices who should be held responsible for the harm done? Held accountable by whom and in what way? To answer these and similar questions, we will need to think about the context of globalization, the coherence of a broad consequentialism, and the practical obstacles facing an ethic for development guided by principles of inclusion, identity, and fair discourse.

---

[6]The concept of complicity has been used to discuss the responsibilities of groups and institutions during the period of apartheid in South Africa. See Truth and Reconciliation Commission (1998). See also Esquith 2009.

Chapter 8
Agricultural Intensification: Some Human Rights Issues

John Otieno Ouko*

Intensification of agriculture has been essential for centuries in meeting the increasing demand of growing world population for food. Over the past few decades, intensification has taken the form of Green Revolution-style farming. However, it seems there are human rights challenges to be faced especially if we look at Green Revolution-style farming in the wider picture of economic globalization.

In this paper I will investigate whether actions that result from Green Revolution-style farming are consistent with human rights and duties. First, I will start by presenting an argument that food is basic human right. Second, I will take a closer look at the idea of consumer rights in the light of the notion that food is basic human right. Finally, I will examine Green Revolution-style farming within the framework of economic globalization.

Food as a Basic Human Right

According to various researchers, including Jeffrey Sachs, Peter Singer, Thomas Pogge, and Rose M. Welch[1] amongst others, hunger now afflicts about two billion people worldwide. Generally there is an agreement among human beings that hunger is a bad thing and that it would be a good thing to alleviate world hunger.

According to Allen Buchanan:

> The dominant contemporary philosophical theories of human rights hold that human rights are grounded in certain common characteristics of human beings. For example, James Nickel, Henry Shue, Amartya Sen, and Martha Nussbaum each ground human rights in common characteristics of human beings- more specifically, in a conception of basic human interest (or capabilities in the cases of Sen and Nussbaum) understood as those interests that must be realized (or those capabilities that an individual must have) if a human being is to have an opportunity to live a decent or minimally good life. (2006)

*Graduate of the University of Nairobi and Ph.D. student in philosophy at Michigan State University, focusing on ethics and development

[1] For more information the following materials may be helpful: *The End of Poverty* by Jeffrey Sachs, *World Poverty and Human Rights* by Thomas Pogge, and the Welch and Graham article (1990).

P.B. Thompson (ed.) *The Ethics of Intensification*,

There is a category of rights called basic rights. According to Henry Shue there are three basic rights namely right to life, right to health and right to subsistence. Shue wrote: "basic rights are everyone's minimum reasonable demands upon the rest of humanity. They are the rational basis for justified demands the denial of which no self –respecting person can reasonably be expected to accept" (Shue 1980). The protection of basic rights is important and basic because their protection is essential for the enjoyment of other rights. The three basic rights are human rights because they are rights of the individual as a member of the human species, not merely as a citizen of a given state. They are also universal because their preservation is essential for the preservation of the value of humanity in the universe, i.e., necessary for ensuring a humanized life on Earth. The basic rights are also the "inherent rights of persons." They are "inherent" because, for any individual to be able to exercise the function of person, she or he needs at least the fulfillment of these rights as a necessary condition. To deny them is to deny individuals the very minimum for the function of being a person; it is to deny the right to self-preservation (Oruka 1997).

Self preservation, as we are aware, is actually the object of one's right to life. And to have a right to life for an individual entails satisfying the needs which sustain existence. And one of the most basic of these needs is subsistence. Subsistence is a necessary basic requirement for human survival and for the exercise of functions of a person. Subsistence entails having food or being free from hunger. But as Rose M. Welch et al. and many other researchers have noted hunger now afflicts two billion people worldwide, resulting into poor health, increased rate of chronic diseases (coronary heart disease, cancer, stroke, and diabetes), and permanent impairment of the cognitive abilities of infant born to micronutrient deficient mothers, and high rates of mortality (Welch and Graham 1990).

Owing to the importance of the right to subsistence, right to food has indeed been acknowledged even in the international instruments. For instance, Article 25 (1) of that United Nations Declaration of Human Rights reads: "Everyone has the right to a standard of living adequate for the health and well-being of himself and of his family, including food…" The declaration further demanded that member nations should, "by progressive measures, national and international … secure their universal and effective recognition and observance."

As Asbjorn Eide and others (1984) have noted,

> Within the collection of covenants, declarations, and charters that make up the body of international human rights law, there are a number of articles in which it is established that "everyone" has a right to food. The fact that the right to be free from hunger and malnutrition- to have enough food to ensure physical and mental wellbeing- is written into an established system of human rights, provides those concerned with hunger and development with the opportunity to bring to bear on those issues a new kind of leverage.

"Food as a human right" may be an unusual way of approaching the problem of food. For food is normally regarded as a basic need, and from this perspective, people and governments are expected to naturally attune their productive capabilities to its satisfaction. To assert it as a right sensitizes us to certain realities that are otherwise not evoked when food is merely described as a basic need. As Martha Nussbaum has noted, when we talk about human rights "it tells people right away that we are dealing with an especially urgent set of functions, backed by a sense of justified claim that all humans have to such things, by virtue of being human" (2003).

Just saying that food is a human right doesn't put food on the table. Turning the right to food into a reality for present and future generations is a hard work. It takes a series of political decisions, technological innovations, social cooperation, and individual actions to make it happen. Here is where Green Revolution-style farming comes in. It is hoped it will lead to improved crops (in developing countries) which would provide cheaper and more reliable food locally, reduce mortality and malnourishment, and perhaps assist development in other ways. If this is the case then Green-Revolution style farming is something that ought to be promoted. But there are concerns about it which I now proceed to discuss.

Securing Food Needs and the Question of Consumer Rights

To begin with, Green Revolution-style farming generally involves genetic modification of crops to increase yields. This is necessary if we have to increase food production to feed those who are starving. However, one of the concerns about Green Revolution-style farming that has been raised is that it is not geared towards production of indigenous foods. Therefore it may lead to the violation of the rights of those who can afford yet are unable to access these indigenous foods. The question is: How will they choose not to eat GM food if alternatives are not available? There is doubt as to whether people's rights will be maintained, for example, the right to freedom of choice as consumers. And this seems to be a legitimate concern.

One may rebut the above right claim by arguing that a right to have a product made available even when the market would not otherwise have supplied it presents grave difficulties (Nuffield Council of Bioethics 1999). The question here is what matters most between having indigenous foods in the market so that some people's right to indigenous food can be fulfilled and fulfillment of the right to subsistence of everyone including the poor. The argument that the right to have variety of indigenous foods in the market can be overridden by the right to subsistence of the world poor sounds plausible. If through intensification we can produce food to feed the starving world population then we should not completely refrain from intensifying because it will lead to non-fulfillment of some people's right to have indigenous food available to them. The right of the consumer to have variety to choose from can therefore be constrained by the need to produce large quantities of a few varieties through genetic modification. Still the rights of the consumer to choose not to consume GM food and perhaps to have non-GM foods kept available in spite of market pressures tending in the opposite direction remains a legitimate concern. As Susantha Goonatilake has pointed out, "preserving the present diversity of food sources- as well as regaining some of what is already lost- would reduce humanity's vulnerability to the brittle fragility of the present food supply system…the cry for the right to cultural identity in food is an appropriate one" (Eide et al. 1984).

As we embrace Neo-Green Revolution it is important to ensure that we retain food variety for different social groups but this should not be done at the expense of non fulfillment of the right to subsistence for all human beings especially if it is not the case that there are some nutrients which are vital for human body which can only be found in indigenous foods. The reasons that can be given for the right to have indigenous food variety available should be considered in light of the right of the world poor to subsistence.

When we consider the introduction of a new technology, such as that related to GM crops, there is another aspect of consumer right that comes to mind, the right to be informed about the food one is eating. As Nuffield Council of Bioethics has reported:

> Farmers, food processors and government of the US have resisted requests, mostly from Europe, to accept mandatory labeling requirements which could allow consumers to avoid GM products, especially when the two alternatives are not substantially different. Such labeling requirements protect choice as a value independent of consumer safety, but at a cost… US producers have viewed elaborate labeling requirements, imposed to allow consumers to choose not to consume GM foods, no matter what their reasons are, as restraints on trade under another name. Choice raises the issues of rights. Considered as a matter of rights, the 'balance' to be struck is that between the expectation of commercial firms that they will be able to operate in a predictable environment and the right of the consumer to choose what he or she eats. (1999)

From the preceding, it seems that farmers, processors and the US government's main concern is to have a stable market for their products. This focus is quite removed from what the primary aim of producing GM foods ought to be: to feed the ever growing world population. Even though profit making is an important aspect of the business, it is still important that consumers be allowed to know whether the food they are consuming is genetically modified or not. The consumer should be allowed to choose. Perhaps the price of the indigenous foods should be raised if it turns out that the cost of producing them is higher than that of producing GM foods or if it turns out that they are less economical to produce.

In brief, my position is that consumers have the right to know whether the food they are consuming is GM or not. For this reason labeling should be mandatory even if this would enable some consumers to avoid GM products. Our primary focus should be to meet everyone's right to subsistence and to do this without compromising consumer safety. If the right of the hungry to subsistence cannot be met if plenty of indigenous foods is to be available in the market for those who want them even if GM foods and indigenous foods are not substantially different, then the right of the poor to subsistence outweighs the right of the affluent to have varieties of indigenous foods available at all cost.

My position, therefore, is that Green Revolution kind of agriculture is a good to be promoted especially since it promises to feed the world hungry. The question is whether this promise can be realized or whether it is a pipe dream, and what makes it a pipe dream if it is. To look at this we now proceed to the question of institutional constraints on agricultural intensification and the idea of patent rights.

Green Revolution-Style Farming and Economic Globalization

Capitalistic Activities of Multinational Chemical-cum-Biotechnology Companies

As Deane Curtin maintains, what Green Revolution-style farming reveals is an attempt to radically transform the third world's agriculture from subsistence farming to mechanized agriculture that focuses on export crops. In short it seeks to globalize world agriculture into a capitalist model of food production. Very often, local subsistence is sacrificed for corporate benefit (Curtin 2005). Curtin's observation notwithstanding, some of the fundamental questions are: who is in charge of the process? Is its purported objective of feeding the world poor being realized? A closer look at this method of agriculture reveals that developed countries, multinational corporations and sometimes global institutions like the World Trade Organization and some other private organizations are behind the process. But it is questionable whether the prime objective is to feed the world's hungry.

To begin with, it is apparent that some of the driving forces of Green-Revolution style farming are multinational companies (MTCs). Monsanto, Dupont and Novartis are some of the multinational chemical-cum-biotechnology companies that have been playing a major part in Green Revolution-style farming even in developing countries; and they have been promoting the idea that genetic engineering will boost crop yields and feed the hungry (Rosset 2000). And as Curtin has noted, agrochemical companies, including the ones mentioned above, are now engineering seeds to respond (or not respond) only to their particular commercial brand of fertilizer, herbicide, or pesticide (Curtin 2005). Additionally, Green Revolution hybrid seeds (engineered by these companies) cannot be replanted since they are not self-pollinating. In other words, farmers have to purchase seeds every year. Moreover, green-revolution farming style involves plant monoculture. This makes high yielding varieties especially susceptible to pests and diseases, hence its dependence on pesticides. In contrast to agriculture based on the principle of recycling inputs, the green revolution depends on external sources for seeds, chemicals, and machinery (Cleaver 1972). As the Nuffield Council of Bioethics has noted, "…resource poor farmers are not able to replant the seeds every year. Such a strategy is not possible with today's commercial hybrid varieties" (ibid.). What this does mean is that resource poor farmers may be ultimately locked out of farming if they can't afford to purchase the seeds and other inputs every year.

As Peter Rosset has pointed out, more recently developed seeds may produce higher yields even without manufactured inputs, but the best results require the right amount of chemical fertilizer, pesticides, and water. So as the new seeds spread, petrochemicals become part of farming. In India, adoption of the new seeds has been accompanied by six-fold rise in fertilizer use per acre. Yet the quantity of agricultural production per ton of fertilizer used in India dropped by two-thirds during the Green Revolution. In fact, over the past 30 years the annual growth of fertilizer use on Asian rice has been 3 to 40 times than the growth of yields.... Because farming methods that depend heavily on chemical fertilizers do not maintains soil's natural fertility and because pesticides generate resistant pests, farmers need ever more fertilizers and pesticides just to achieve the same results. As Vandana Shiva recently reported, "Indian peasants who have been made poor and pushed into debt over the past decade to create markets for costly seeds and agrichemicals through economic globalization are ending their lives in thousands" (2005).

The above being the situation, the agrochemical companies' tendency to engineer seeds to respond (or not respond) only to their particular commercial brand of fertilizer, herbicide, or pesticide is suspect. As Cleaver maintains "the integration of the agricultural sector into capitalist market is achieved through the adoption of the new technological package which itself makes the peasant producer more dependent on the market. The farmer must now buy the manufactured inputs and sell part of his crop for cash in order to be able to buy next year's inputs" (Cleaver 1972). Due to the capitalistic nature of these multinational agrochemical companies their focus may not after all be on improving food production but on profit maximization even at the exclusion of peasant producers.

A recent report from Ethiopia is a testimony to the above claim. There are those farmers who cannot afford seeds and fertilizers. As has been reported by the Human Rights Watch:

> The Ethiopian government distributes fertilizer on credit to poor farmers throughout the country in an effort to boost their productivity. Many farmers welcome the help, but it comes at a very steep cost. The fertilizer, sold by one of the multinational agrochemical companies, is so expensive that farmers often find it difficult to repay their debts. Local officials routinely throw farmers into prison for days or weeks at a time if they fail to make debt payment. (2005)

The situation is not different in Kenya where use of agricultural inputs notably fertilizer has declined following the implementation of structural adjustment and agricultural reform programs in the 1990s. Under this program the government is not allowed to subsidize the agricultural sector. Agricultural inputs including fertilizer are therefore very expensive (Mkangi 1993).

Perhaps it is only governments of developed nations that can come in to remedy the situation since, after all, they have repeatedly come together to pledge their commitment to end world hunger. But there are indications that governments of developed countries are not committed to ending world hunger. As Theodore Schultz has noted:

> While governments in developed countries tend to give agricultural scientists within their own countries substantial authority in selecting research projects, when these same governments provide funds for agricultural research in the less-developed countries they tend to sponsor narrow, inefficient mission-oriented research. Moreover, the support is subject to the vicissitudes of foreign policy; it therefore lacks the long-run stability that is absolutely necessary in developing effective organized agricultural research. (1972)

There are many areas of research with potential benefits to the poor that are not being carried out by the private sector. As Curtin puts it, "the fact that high yielding varieties were chosen for development, while programs for open-pollinated varieties were forcibly shut down, reflects the class interests in green revolution. High yielding varieties were engineered to be a privately owned commodity" (2005). Research remains to be done, for instance, on drought-resistant millet, with a much less certain commercial payoff, but a high food security impact among poor Africans.

In cases where the public sector has recognized its role and decided to invest in such "orphaned areas" its work is hampered by inability to use freely basic but proprietary knowledge generated by private research. Rights at stake also include the right of the citizens of developing countries to have their interests considered in the policy decisions of the regulators, researchers and agrochemical companies in the developed world. As a whole, as Nuffield Council on Bioethics has pointed out:

> "The role of appropriate national and international infrastructures in supporting the world's poor is well illustrated by the controversy over apomixis genes. It may soon be possible to develop plant varieties that can produce seeds without sexual fertilization...This would revolutionize plant breeding by allowing any desired variety, including hybrids, to breed, thus permitting plant breeders to more readily develop locally adapted genetically diverse varieties. Moreover, resource- poor farmers would be able to replant the seed every year. Such strategy is not possible with today's commercial hybrid varieties...the current trend towards the consolidation of GM plant technology ownership into a relatively small number of companies may severely restrict access to affordable apomixis technology" (1999).

From what we have seen multinational agrichemical companies, whose trade activities are of course endorsed by global institutions like the World Trade Organization (WTO), will most certainly prioritize profit maximization and not ending world hunger. This has been the trend. The activities of these multinationals can be seen more clearly when it comes to the issue of patent rights for the seeds, to which we now proceed.

Patent Rights and Institutional Order

Under Article 34 of the World Trade Organization's (WTO) Agreement on Trade and Related Aspects of Intellectual Property (TRIPS), private sector companies are allowed to have patents for the products that they develop and choose to sell. Supporters of patenting point out that if the private sector is to mobilize and invest large sums of money in agribiotechnology, it must protect and recoup what it has put in or invested. This return on investment they get in form of loyalties paid by consumers of their products. According to the same Article 34 of the WTO's TRIPS, if one variety, say of rice, has multiple patented genes, farmers (consumers), whether small holders or not, are supposed to pay multiple loyalties for the multiple genes. For example if you take a variety of rice like Golden Rice which has 70 patented genes which belong to 32 patent holders (owners), farmers, whether small holders or not, are supposed to pay (the 32 patent holders) loyalties for the 70 patented genes.

Whereas it is a good idea to acknowledge and appreciate the work of plant breeders, it seems that the idea of patent rights is exaggerated. It goes beyond merely recouping what has been invested. As Shiva wrote, "when seeds are patented and peasants will pay $1 trillion in royalties, they will be $1 trillion poorer…with privatization of essential services and an unfair globalization imposed through WTO, the poor are being made poorer" (2005).

Moreover, when a small holder plants a variety with multiple patented genes as in the case of Golden Rice, according to the same Article 34 of the WTO's TRIPS, it would subject the non-GM varieties biologically affected by the patented variety to the licensing requirements of the GM crop. As Masood has noted (1998), "the bizarre twist that this adds to the whole process of globalization is that smallholder farmers could unwittingly become patent infringers."

As a whole, patenting leads to monopolization of knowledge (by preventing the national agricultural research systems of developing countries from using the same basic scientific processes to develop products that would benefit poor people), restricts access to germplasm, controls the selectivity in research focus, and increases marginalization of the majority of the world's population. As a result, countries no longer control their food security. Corporations do. This aggravates starvation of millions of people (Curtin 2005). The question one might ask is this: is this the way to help farmers in developing countries? We may conclude with Tewolde Egziabher that "if genetic engineering is to help farmers in the south, it has to be freed from the patent knots that is making it a global trawling net" (2004).

Conclusions

The pre-science agriculture as it still prevails in many parts of the world is at the mercy of the vicissitudes and niggardliness of nature. Raw land and brute labor are not sufficient to dispel the ancient fear of starvation (Schultz 1972). Therefore, Green Revolution-style farming should not be abandoned as such. If poorer countries are excluded from adopting GM seeds, their domestic food supplies may be deprived of potential improvements. As a result we may end up with more people starving.

Food being a basic human right, in fulfillment of their imperfect obligation, developed countries should assist poor countries to do more research with potential benefit to the poor (i.e., to come up with crops whose seeds can be replanted every year) including research on drought-resistant crops. This will be a far much better idea than giving food aid during famine or drought since it will help the poor to be more self reliant in terms of food production.

Finally, multinational chemical-cum-biotechnology companies and global institutions like the WTO should also ensure that they fulfill their negative duty not to harm the poor farmers in developing countries. As Pogge has observed (2002), the

WTO is effectively controlled by only a handful of powerful developed countries and the policies implemented by the WTO have profound impact on the developing world. He contends that global institutions have a causal relationship to global poverty. A case in point is the World Trade Organization's Agreement on Trade and Related Aspects of Intellectual Property (TRIPS). As we saw, TRIPS is designed to benefit the rich and poor farmers cannot survive under it.

Chapter 9
Environmental Ethics and Agricultural Intensification

Clare Palmer*

Introduction

In his book *The Spirit of the Soil*, Paul Thompson notes that "environmental philosophers have devoted little attention to agriculture" (Thompson 1995). He suggests, plausibly, that this lack of attention is, in part at least, because environmental ethicists have tended to focus on the value of natural areas in an untransformed state – and that thus they feel discomfort with the intrinsically transformative nature of agricultural production. While some of the issues considered by environmental ethicists are relevant to an environmental philosophy of agriculture, he suggests, "they are sufficiently disengaged from agricultural problems that their applicability is far from obvious."

In this paper, I will attempt to bring some of the leading approaches to environmental ethics to bear on agricultural issues – specifically on agricultural intensification – in order to move a little closer to the kind of engagement Thompson recommends. To do this, I will first briefly outline key features of these leading approaches to environmental ethics. I will then suggest how such approaches might work out in an agricultural context, and in particular how they might regard further agricultural intensification. (I will mostly concentrate on possible future intensification, rather than considering the questions for environmental ethics raised by past intensification.) It will become clear from this discussion that some of the fundamental value positions central to these leading approaches to environmental ethics will be problematic to many working in agriculture and thinking about agricultural policy. And even if one accepts some of these values, difficulties clearly remain about how they should be weighed up in decision-making against one another, and against other kinds of values. So in the final part of the paper, I consider whether broadly "pragmatic" approaches to environmental ethics might be more helpful in thinking about agricultural intensification.

---

* Associate Professor in Philosophy and Environmental Studies at Washington University in St. Louis and president of the International Society for Environmental Ethics

P.B. Thompson (ed.) *The Ethics of Intensification*,

All the environmental ethics positions I consider here hold key values in common with one another, and with values widely held outside the discipline of environmental ethics. All maintain that human life and well-being are valuable. All assume that benefits to humans have flowed and do flow from agricultural intensification, and that without some kinds of agricultural intensification human society as we know it would have been impossible. However, many leading approaches to environmental ethics – while accepting this – also take *other* concerns into account. These concerns run alongside those relating directly to production and human welfare, and cluster around what is considered to be of value in the environment *aside* from what is directly useful to humans.

Key Terms and Approaches in Environmental Ethics

To make sense of what's going on in discussions in environmental ethics, it's important that I begin by making a couple of key distinctions. The first concerns different ethical theories that can be adopted within environmental ethics. In brief, two kinds of approaches dominate ethical theory. (A third, which I'll mention later, focuses on virtues and is growing in significance.) One ethical theory is called *consequentialism*. Consequentialists argue that ethics is concerned with bringing about the best possible state of affairs in the world. On most versions of this view, therefore, we should decide what to do by looking to the future and thinking about the consequences of our actions. Adherents to an alternative ethical theory – usually called a *deontological* or *duty-oriented* approach, argue in contrast that consequences are not all that's relevant when we make moral decisions. So, for instance, we might have made a prior commitment such as a promise; duty oriented theorists will usually argue that promises should not be broken even if doing so would bring about the best consequences. Duty-oriented theorists also argue that there are duties – such as a duty not to kill – that should be respected even if, in some cases, killing would bring about a better state of affairs (an example might be the assassination of a murderous dictator). The difference between these ethical theories has been important for debates in environmental ethics.

The second distinction to make here – one about key terms – is that between instrumental and intrinsic value. *Instrumental value* is usually used to mean value assigned to something because of its usefulness as a means to an end. Water, for instance, is of instrumental value to humans because it helps in achieving another goal – that is, remaining alive. But this does not seem to be the case with all kinds of value – for instance, being alive itself. We do not (usually) value our lives for any reason beyond themselves; we do not (usually) regard preserving our life as a means to an end, but rather as an end in itself. Value of *this* sort is non-instrumental value – often called *intrinsic value*.[1] These two kinds of value are not mutually

[1] Intrinsic value can be used in a number of different ways. This explanation is intended to make clear that by intrinsic value I am referring to *non-instrumental* value, not to the alternative possible uses as *objective* (as opposed to subjective) value nor to *non-relational* value (the sense usually attributed to G.E. Moore). See O'Neill (1993).

exclusive: something can be simultaneously of instrumental value in one way, and intrinsic value in another. An apple tree could be of instrumental value as a provider of fruit while of intrinsic value as a living organism. A further point should be made here though: to say (for instance) that an organism, a system or a process has intrinsic value is not necessarily to say that it has *rights*. The term 'rights' is usually used in the context of beings with particularly high moral and political status; the sphere of moral concern can be much broader than that encompassed by rights alone.

One central focus of debate in environmental ethics concerns what (if anything) is of intrinsic value in this way in the environment. A number of different answers have been proposed. I'll begin by outlining some of these answers, and then think about how these might relate to agricultural intensification. Of course, since agricultural intensification has taken and can take many forms in many different places, cultures and ecosystems, here I'll be able to provide only a fairly perfunctory account of a very complex area.

Possible Intrinsic Values in Environmental Ethics

Sentience and Welfare

For some environmental ethicists, and certainly for those who are particularly concerned with the welfare of individual animals, something we may broadly tag "sentience" is thought of as being of intrinsic value – though what is meant by sentience is much disputed.[2] For simplicity's sake, I'll use Warren's definition of sentience: "The capacity to have conscious experience, usually including the capacity to have pain and pleasure" (Warren 2000).[3] Further, the way in which sentience might be of intrinsic value is disputed. Some philosophers argue that it is the possession of the *capacity* for sentience that is significant (they might argue that this capacity forms the basis, or one of the bases, for the attribution of *rights*); for others it is not so much the capacity itself as the resulting *sentient experiences* that are of intrinsic value – or disvalue. On this account, sentient experiences of pain are of intrinsic disvalue, and sentient experiences of pleasure are of intrinsic value. However the value of sentience is interpreted, this approach is primarily concerned with the *welfare* or *well-being* of humans, other mammals, and – though this is much debated – fish, reptiles and a few other organisms believed to be sentient (such as octopi).

Both consequentialists and duty-oriented theorists may adopt sentience-centered approaches. For consequentialists, concerns about the consequences of our actions for the wellbeing of humans can be extended to include relevant consequences

[2] Some argue that animal-centered positions should be excluded from environmental ethics. This is not the view that I take, but here animals are included in environmental ethics for simplicity's sake; no stronger commitment is intended by it.

[3] Even this is problematic, since what is meant by "conscious" is hotly contested.

(pains and pleasures, for example) for animals as well. Duty-oriented theorists may argue that because animals are sentient, we have *prima facie* duties not to harm them. Some ethicists in this group – such as the animal rights theorist Tom Regan – take a very strong position here. This strong position rejects the idea of animal agriculture altogether, however intensive or otherwise it might be, on the grounds that animal agriculture *in principle* treats animals as instruments, solely as means to a human end, and thereby overrides their rights, including their right to life. However since these arguments don't relate to intensification *per se*, I won't pursue them further here.

Life

Being alive is, in some accounts of environmental ethics, both necessary and sufficient for a being to have intrinsic value. Views of this kind have a long history – for instance, one version was put forward in Albert Schweitzer's advocacy of the principle of *reverence for life* in the 1920s.[4] More recently, such a view was proposed in Kenneth Goodpaster's much reprinted paper "On Being Morally Considerable," where Goodpaster argues that "nothing short of the condition of being alive seems to me a plausible and nonarbitrary criterion" (of possessing moral status).[5] A more developed version of this view was proposed by Paul Taylor in his book *Respect for Nature* (1986). Taylor argues that all organisms pursue their own good in their own way, and that they are thus ends in themselves. Their *telos*, or *end*, gives each individual organism *inherent worth* (which Taylor uses in a way similar to my use of intrinsic value) and this inherent worth is equally possessed by all living organisms, since all have a good of their own.[6] Indeed, Taylor maintains a controversial principle of *biocentric impartiality*, arguing that, in principle, all living individuals have the same value since inherent worth in some way flows from *being alive*. Taylor's book focuses on wild organisms rather than the organisms of what he calls "bioculture" – in which he includes agriculture (Taylor 1986). Yet, given the criteria he adopts, his case seems to work for all living things, whether wild or not; all living organisms are, for Taylor, of equal inherent worth. I'll call approaches of the kind adopted by Goodpaster and Taylor "respect for life" views.

Species

The idea that species are intrinsically valuable is a widely held view, both within and outside philosophical environmental ethics. Principle 1a of the Earth Charter,

---

[4] Though Schweitzer also considered some things (such as crystals) that we do not customarily think of as "living" as expressing the will to live.

[5] See Goodpaster's paper for a fuller defense of the position.

[6] Taylor's argument is, of course, more complex than this suggests, but this suffices in summary.

for instance, instructs us to: "Recognize that…every form of life has value regardless of its worth to human beings."[7] This sounds like an affirmation of intrinsic value. Within environmental ethics, a number of different arguments have been made for the intrinsic value of species. Some have argued that a species is a kind of super-organism (it evolves, it can reproduce, it can die) and should be valued in the way one values any other living being (making a species into a special case of living organism as in (b) above). Perhaps the best known argument here, though, has been made by the environmental philosopher Holmes Rolston. He maintains that, while individual organisms do have intrinsic value, species are the "life-lines" to which respect is owed. Individuals are the tokens of species, the type; it would be odd to value the tokens without valuing the type from which they emerge and which they instantiate. Rolston maintains (2003) that "if it makes any sense to claim that one ought not to kill individuals without justification, it makes more sense to claim that one ought not to extinguish species lines without extraordinary justification. This is a kind of superkilling."[8] Rolston's concern here seems to be not so much for the species as a kind of constant 'form' but more for the continuance of the evolutionary processes that allow for the instantiation of individuals of the species (and also for evolutionary adaptation). As such, this moves close to valuing *biodiversity*, where biodiversity is understood in relation to the continuance of evolutionary processes of adaptation.

Ecosystems/Ecological Processes or Qualities

Sometimes it is argued that an ethic is inappropriately called *environmental* if its focus is on the capacities of individual organisms. Rather, it is suggested, environmental ethics should focus on ecosystems. It is these, or particular states they can be in – such as *healthiness, diversity, stability, integrity, complexity* or sometimes *wildness*, that carry intrinsic value. Holmes Rolston (1994) argues for a particular kind of value – *systemic value* – which he describes as "a creative potential that steadily becomes actual." Not to accept the existence of this kind of systemic value is, for Rolston, to propose a truncated vision of the world: "Valuing the products but not the system able to produce these products is like finding a goose that lays golden eggs and valuing the eggs but not the goose."

One important question in the context of agricultural intensification is whether such an ecosystem ethic can include land used for agriculture. Obviously, if

[7] This is actually open to two interpretations. It may mean regardless to its *use* to human beings, which would correspond to my sense of intrinsic; or it might mean "regardless of whether humans value it or not" which may be a claim that species have objective value, independent of any valuer. The latter claim, if intended, would be much more problematic than the former. The text of the Earth Charter can be found online at http://www.earthcharter.org.

[8] I quote here from Rolston's essay in *The Blackwell Companion to Philosophy* (2nd ed.). His views on species, though, are developed at more length in Rolston (1985).

untouched wildness is seen as a requirement for ecosystem value to exist, then agricultural land cannot be valuable in this sense. But other understandings of ecosystem value do not exclude agricultural land. If one argues that an area of agricultural land is healthy or flourishing, for instance, by particular ecological criteria, and these criteria also carry intrinsic value, then agricultural land can carry intrinsic value. The best known proposal for a land ethic, from the work of the forester Aldo Leopold (1949), claims that "A thing is right when it tends to preserve the integrity, stability and beauty of the biotic community. It is wrong when it tends otherwise." It has been widely argued – by Paul Thompson, J. Baird Callicott and others – that Leopold did not intend to exclude agricultural land in making this claim.

One more general point to be made here: these different values are not necessarily mutually exclusive. Certainly, some environmental ethicists argue for only one sort of intrinsic value, to which all other values are instrumental. For instance, a number of ethicists consider only positive sentient experiences to be intrinsically valuable, and suggest that ecosystem or species values are instrumentally valuable only in as much as they underpin such valuable sentient experiences. On the other hand, a number of environmental ethicists are value *pluralists*, maintaining that there are different kinds of values, all of which are important, and that these values may sometimes come into conflict.[9] For these ethicists the problem is how different values are to be prioritized, and whether (and in what ways) it might be said that values are incommensurable. I'll consider value pluralism at more length later in the paper. First, I'll move on to relate these different traditions in environmental ethics to issues raised by agricultural intensification.

Intrinsic Values, Environmental Ethics and Agricultural Intensification

Agricultural Intensification and Sentience/Welfare

Agricultural intensification has already had a huge impact on sentient animals, both directly (in terms of the intensification of animal agriculture itself) and indirectly (in terms of the way arable agricultural intensification has affected both wild and domesticated animals). Intensification of animal agriculture looks set to continue in at least three major ways:

(a) The spread of existing intensive methods to countries where intensification in animal agriculture has not previously been widely practiced.
(b) The extension of intensive methods to animal species not previously farmed intensively, or not farmed intensively on a large scale.

[9] There are other forms of pluralism in environmental ethics too, and there are also different strengths of pluralism. For one kind of account, see Wenz (1993).

(c) The development of biotechnology.

In addition:

(d) Intensification of arable agriculture, if it continues, is likely to have further impacts on wild sentient animal life. But it is worth noting that increased *extensification* of agriculture – where agriculture expands into previously wild land – is likely to cause more of an impact on wildlife (as well as on wild ecosystems and endangered species) than increased agricultural *intensification*.

Further intensification of farming methods may worsen agricultural animal welfare, improve welfare, or worsen some aspects of welfare (such as the ability to move without restriction) whilst improving others (such as resistance to disease). Clearly, some existing forms of intensive animal agriculture do restrict movement and generate chronic ill-health and pain among animals – as in the case of broiler chickens (Webster 1994; CIWF 2003). For an ethicist concerned with the welfare of sentient animals, these kinds of intensive systems are ethically problematic, and their expansion by export to other countries is likewise problematic. A consequentialist would attempt to weigh up whether the benefit to human welfare from the introduction of more intensive animal agriculture to poorer human communities (assuming there to be a benefit) would outweigh the cost to the animals. But duty-oriented ethicists who consider that sentient beings are of intrinsic value might argue that welfare gains to humans should not be stacked up against the pain and stress inflicted on sentient animals: we have duties not to harm them in these ways even if there are some human benefits from doing so. By such reasoning, if it is wrong to create and keep animals with poor welfare, then benefit to humans cannot right the wrong.

A number of particularly interesting, but complex, ethical issues are raised by the prospect of further intensification in animal agriculture, which I can mention only briefly here:

(a) *Animal numbers*: Intensification may have an impact on total numbers of agricultural animals in existence. Intensive methods may mean that fewer animals in total need to be kept to produce the same quantity of animal products. For instance, artificial insemination means that the need to keep male animals has fallen; and there has been a 50–150% increase in efficiency of extraction of animal products from animals (Seidel 1998). Intensification, then, might result in fewer animals being kept than at present, or in fewer animals being kept than would otherwise be the case in more extensive systems. Is this good? Responding to this question depends on difficult issues about whether it is good (other things being equal) to bring sentient lives into existence, and how far this depends on the quality of the sentient lives concerned.[10] Is it better (for instance) to have fewer intensively farmed animals living very painful and stressful lives, or greater numbers of less-intensively farmed animals living less

[10] Peter Singer considers aspects of this question in his discussion of what he calls the "prior existence" view of utilitarianism (influenced by Parfit). See Singer (1993: 103–109).

painful and less stressful lives (though perhaps still not lives we might call "good" lives)? Is a creature harmed by being brought into existence if it is predictable that its experience of living will be predominantly unpleasant or painful? (See Parfit 1984).

(b) *Value conflicts*: There are cases where intensive farming may be detrimental to agricultural animal welfare, but may promote wild animal welfare and other environmental values. For instance, changes to EU agricultural subsidies make it possible that UK cattle farming will intensify, resulting in cattle being kept indoors year-round.[11] This move is likely to have negative welfare implications for cattle, in terms of their freedom of movement and their potential to develop painful conditions such as lameness. However, one of the aims of the intensification of cattle farming is to free up pasture land to promote land management for wildlife, without reducing overall cattle numbers. This is likely to increase the numbers and improve the welfare of wild animals as well as promoting ecosystem values and perhaps species values. Here, then, conflict might be generated *between* ethical positions. Those who focus on species or ecosystem values may regard this intensification of cattle farming as good (although this will depend on the environmental and wildlife impacts of silage effluents, slurry and the dirty water produced from yard-washing, as well as whether the existing ecology that has developed including the grazing cattle is viewed as valuable). Even those primarily concerned with animal welfare would have to consider the positive effects of taking the cattle off the land for the welfare of wild animals. Disagreement between environmental ethicists who focus on the value of systems/species, and those who focus on the value of individual welfare is not uncommon (for instance over hunting and culling). Occasions where intensification may worsen the welfare of agriculture animals but promote other, environmental values may become increasingly common and contentious in future years.

(c) *Biotechnology*: While some forms of intensification change the *external circumstances* in which animals live, other forms of biotechnology change their *bodies or natures internally*. Some selective breeding – where broiler chickens are unable to carry their own weight, resulting in leg deformities; or where cows are unable to give birth to calves that have been selected for "double-muscling" and require repeated caesarean sections – causes a decline in welfare (Webster 1994; Food Ethics Council 2001). Biotechnology developments could accelerate welfare-negative processes like these. However, other biotechnological developments promise improvements in animal welfare: for instance, by the insertion of genes for protection from disease, and the modification of animals so that the need for surgical castration and dehorning is removed. In a slightly more distant but not implausible future it may become possible to breed animals with less desire to move around or that are less sensitive to pain – the production of "placid, fat, happy bulls" or even "deaf, blind, legless microcephalic lumps" (Cooper 1998). If the concern of ethicists is purely with sentient animal

[11] BBC Radio 4, *Costing the Earth*, 4th December 2003.

welfare, such developments should be welcome. And certainly some welfare/sentience centered ethicists think so. Bernard Rollin (1997), an important figure in the animal welfare field, maintains that if (for instance) one could engineer a chicken that achieved satisfaction by laying an egg in a cage rather than a nest, and chickens are housed in cages, then a cage-satisfied chicken is living a better life than a nest-satisfied chicken. For Rollin, if an animal's nature is altered so that the dictates of its nature can be more nearly satisfied by the circumstances in which it is born to live, then this would be a morally good use of biotechnology. However, some welfare-centered ethicists are concerned about biotechnology even if it promises welfare improvement, and are troubled particularly about the idea of breeding insensitive animals. Accounting for this worry has led to discussion about respect for animal integrity, though this moves beyond the sphere of welfare and has yet to be thoroughly worked through.[12]

Agricultural Intensification and Respect for Life Views

Respect for life views are usually (though not necessarily nor exclusively) duty-oriented. They tend to maintain that once a life has come into being, the continuance of that life should be respected; further – other things being equal – the life should be permitted to flourish in its species-specific way and not be harmed (there are different views about whether such lives are also owed assistance). Respect for the intrinsic value of a life, seen from this perspective, cannot be traded off in an ethical cost-benefit analysis aiming at the best consequences. Neither, on such an account, are lives replaceable. That is to say, the creation of a new life could not compensate for the taking of another. If a life is taken, the failure in "respect for life" has occurred whether a new life is created or not. (It should be noted that this does not mean that the production of *extra* valuable lives is necessarily a good thing, since valuable lives are not usually thought of as being the kinds of things, from a duty-oriented perspective, that can be added together to make a larger and better sum.)

On this basis, all forms of agriculture might seem problematic and disrespectful since (with the exception, perhaps, of harvesting fruits and nuts) agriculture inevitably involves taking life and curbing flourishing. Certainly, ethicists writing from this perspective tend to regard agriculture as an unfortunate necessity, thus illustrating exactly the difficulties that can be involved in bringing environmental ethics and agriculture together. Taylor (1986), for instance, characterizes agriculture as displaying "absolute, unconditioned power" over nature, power with which he is clearly uncomfortable. Nonetheless, he does accept the necessity of agriculture for the continuance of human society; so his concern here must be to ameliorate its

[12] See Holland (1997) for a preliminary discussion; see also Rutgers and Heejer (1999) and Vorstenbosch (1993).

effects in terms of disrespecting life. (Taylor works through a long series of "priority principles" intended to assist in making decisions of this kind.)

Working from "respect for life" alone, it is difficult to justify a distinction between the value of different *kinds* of organismic life – for instance, the lives of wild organisms, domesticated agricultural organisms or genetically modified organisms. Whatever their origin and relation to humans, they are, essentially, all *alive*. So it is not obvious, for instance, on what grounds someone adhering to such a position might object to biotechnological changes in the creation of living organisms for agricultural purposes. It might be argued that the *attitude* underpinning, e.g., agricultural biotechnology is a disrespectful one, stemming from an assumption of human dominance over, rather than respect for, nature. (Such a position, though, would need to explain why biotechnology differed from selective breeding in this regard.) This idea of "disrespectful attitude" seems close to another kind of ethical tradition I haven't considered here: virtue theory (so, being humble in one's interactions with nature might be seen as a virtue: see Hill 1983). Perhaps one route for developing a respect for life approach in the context of agricultural intensification is to consider how (if at all) agricultural intensification might be undertaken with a "respectful attitude". The following points, then, might arise:

(a) Respect for life entails not taking life and allowing life to flourish, except where necessary for human survival. The deliberate creation by selective breeding or by biotechnology of beings that *cannot* flourish (chickens that cannot walk, for instance) would, on these terms, constitute an unacceptable lack of respect. The ethical problem here is not that the organism has been selectively bred or genetically modified in itself, but rather that in undertaking the genetic modification the possibility of the organism living a flourishing life has been deliberately and disrespectfully curtailed. For organisms modified in particular ways, there may be no possibility of flourishing.
(b) Even though creating more lives is not better than there being fewer lives, it might nonetheless be the case that disrespect for many existing lives is worse than disrespect for a few lives. Taylor (1986), for instance, advocates a principle of "minimum wrong"; arguing that since a duty of non-maleficence is owed to every living organism, if one must cause harm "it would be morally unjustified to harm a larger number of organisms than a smaller number of organisms." On this argument there would be grounds for thinking that modes of intensive farming that involved disrespect to fewer lives would be preferable. This might point towards biotechnological help in producing higher yielding individual plants and animals, resulting in the taking of fewer lives to gain the same yield. (But this might be a *reductio* of the position. Indeed, it seems to result in the strange conclusion that even if one took the view that the *attitude* involved in biotechnology showed disrespect for nature, nonetheless the *outcome* of practising it might be that fewer valuable organisms are harmed.)

Much more investigation of this position – somewhat curious in application to agriculture – is obviously possible; but given that it is a minority view, I will move on to other, more widely adopted values in environmental ethics.

Agricultural Intensification and Species Values

The first question to ask here is what effects agricultural intensification has had and might have on species. Certainly crop intensification to date – in the UK at least – has led to a decline in distribution and abundance of some wild species and to the extinction of others. Artificial fertilizers, insecticides and herbicides have all had, in at least some instances, negative impacts on wild species and biodiversity. But this, of course is not a necessary outcome of intensification. Agricultural intensification can be used to produce more output from existing cultivated land – but equally, it can be used to produce the same amount of output from a smaller area of cultivated land. In the latter case, if the land freed up from agriculture was to be restored as relatively wild habitat, the likely consequence would be to assist the survival of threatened species and to increase wild biodiversity.

Agricultural intensification (whether by selective breeding or by biotechnology) might also have an impact on the diversity of *cultivated* plant strains and animal breeds, if not on whole species. Some few strains only might become dominant, while other strains and breeds might become marginal and die out, or become preserved only in seed banks and centers for rare breeds (this process has happened in come recent episodes of intensification). This would mean a decline in cultivated/domesticated genetic diversity in some species. On the other hand biotechnology can produce *new* cultivated breeds and genetic combinations. These observations raise further questions for environmental ethics with a focus on species values. Is there something special about either existing wild species, or traditional cultivated strains and breeds, that means they merit protection? Suppose it became possible to engineer a new species for every species that became extinct by human actions through intensification. Would this mean that no harm had been done, and no value lost, since the overall total number of species in the world would, after all, remain the same?

Some environmental ethicists might argue – rather as in the "respect for life" case – that one species is not replaceable with another. Even if a new species were to be created with its own intrinsic value, thus keeping the total number of species in the world constant, this would not substitute for the loss of value of the extinct species. Species cannot be added up together in this way. (To make this case effectively, though, it would be necessary to identify what it was about a species that made it valuable in this sense.) A second issue for environmental ethicists here concerns an idea of "naturalness". There's a question about whether a species or strain created by humans in a laboratory is of less value (or of a different kind of value) than a species that has evolved without any human intervention. What's particularly interesting here is that for most environmental ethicists, what is intrinsically valuable about species is not based on the existence of the species alone. Rather, a wild species attracts a cluster of other valuable qualities: take, for instance, Rolston's argument for the value of species, based on their evolutionary and historical lineage. This position on species value is able to sustain a distinction between "natural" species and strains, and those created by humans. Other environmental ethicists might also argue (as, for instance, Robert Elliot maintains in his book *Faking Nature*) that naturalness, or wildness, can be valued in itself. That is, the *origin* of something is relevant to its value, such that a humanly restored landscape can never have intrinsic value in the same sense as a wild landscape. If this view is taken, the creation of new, cultivated strains and breeds can be no substitute for the loss of wild ones.

Agricultural Intensification and Ecosystems/Ecosystemic Processes

This observation leads to the fourth kind of intrinsic value I considered earlier – that relating to ecosystems and ecosystemic qualities and processes. How might agricultural intensification impact upon these? Of course, any response again depends on the kind of agricultural intensification at stake, and whether any kind of agricultural land can be thought to have intrinsic ecosystemic value. If one took the view that only wild ecosystems could have this kind of value, then intensification could generate value if it meant that agricultural land-use contracted, and restored wild land expanded, without the more intensive land-use having a negative impact on the surrounding wild ecosystems.[13] However, as Thompson (1995) argues, this is a very limited kind of environmental ethic, saying nothing about agriculture as such, only about how agricultural land might shrink to provide more room for wild, and therefore valuable, land. But if one took the view that particular values of land states – such as 'healthiness' and flourishing – applied to agricultural land as well as to wild land, then one could consider how proposed intensification would contribute to (or detract from) those values. For instance, intensification of agriculture that involved replacing crops previously cultivated with pesticides by genetically modified high-yielding pest-resistant crops might create healthier, more flourishing rural ecosystems. And it seems perfectly plausible to argue that there can be flourishing ecosystems that include agriculture, even intensified agriculture. Even the great advocate of wilderness protection, Holmes Rolston (1994), suggests that there might be a state of "agricultural integrity," where agricultural areas are "managed sustainably" such that their operation "does not disrupt the surrounding natural systems" and they are "enveloped by natural systems." For Rolston, what matters in agriculture is whether ecosystem connections exist, where (for instance) crops have ecological "fit" in natural local cycles such as those of climate and water and

[13] Provided, that is, one took the view that "re-wilded" land could be of value in the same way as "wild" land.

provide support for, rather than disruption to, local ecosystems. Inasmuch as any particular intensification of agriculture would make these ecosystem connections more possible, such a development would, for Rolston, be ethically desirable; where intensification makes this less possible and (for instance) "individual corn plants might be healthy enough but they are just parked there by humans like potted plants on a porch" (ibid.) Rolston would regard the development as ethically undesirable. Of course, this view might nonetheless be regarded as limited: Rolston's concern is still, primarily, for the wild systems with which the agricultural land is bound up, not with the agricultural land itself. If a stronger view were taken on the value of flourishing agricultural ecosystems, then in some instances intensification could contribute to flourishing, and hence to value.

Agriculture and Aesthetic Value

I have so far exclusively focused on intrinsic environmental values that individual organisms, species or ecosystems may be thought to carry. However, there are, of course, other values on which agricultural intensification impacts – in particular, aesthetic value, which I consider here – as well as a much broader spectrum of human ethical concerns, on which I will focus in the next section.

The aesthetic value of landscape – to which Aldo Leopold seems to have been referring in his inclusion of *beauty* as fundamental to a land ethic – is often seen as being of central importance when agricultural practices change. Many objections to agricultural intensification in Britain, for instance, have revolved not so much around issues of animal ethics, respect for life, species and biodiversity protection, or the value of ecosystems, but rather around the impact of such intensification on the appearance of the landscape. The loss of hedgerows, the erection of factory-like sheds for intensive farming of animals, and the creation of larger fields, often generate concern about the aesthetic – particularly the visual – character of the landscape. Sometimes this concern relates to the loss of the traditional look of a landscape (and, in particular, to ideas of appropriate regional and national landscape); sometimes to the view that the landscapes associated with many forms of intensive agriculture are ugly, and undesirable on the basis. Whilst the visual appearance of landscape (or indeed, the existence of other aesthetic qualities, such as particular smells) may seem a trivial concern in comparison with issues of hunger and development, landscape appearance can be fundamental to individuals' and communities' sense of place, belonging and identity – as well as being of central economic significance in the multi-billion dollar international tourist industry. Indeed, the persistence of some forms of agriculture – such as keeping sheep in the English Lake District – appears largely to be explained by the desire to maintain the aesthetic character of the landscape (grazed hills without scrub cover) rather than by any economic justification (other than that of tourism).

The nature of the aesthetic value of agricultural landscapes has been recently debated among environmental aestheticians. Agricultural landscapes, of course – unlike flower gardens, for instance – are not deliberately created as objects of aesthetic interest. No artistic intention is (usually) embodied in them, unlike in a work of art, even if, in retrospect, particular agricultural landscapes become aesthetically cherished. Allen Carlson (2000) argues that when appreciating agricultural landscapes, rather than thinking about artistic intention, one should consider the *function* of the landscape. An understanding of function helps in aesthetic appreciation in the same way as (he suggests) an understanding of scientific ecology helps in aesthetic appreciation of wild nature. For Carlson, this means that the landscapes of intensive agriculture – given that they are appropriate to their function – can be the location of aesthetic appreciation as much as the small-scale landscapes of traditional farming. Taking this view, he maintains, we could come to appreciate the formal beauty of an intensive industrial farm with its "great checkerboard squares of green and gold; vast rectangles of infinitesimally different shades of grey." Other aestheticians disagree. Emily Brady (2006) for instance, argues that the landscapes of traditional farming are more aesthetically appealing than the landscapes of intensive and particularly industrial farming, on the grounds that traditional farming is laid out on a personal, human scale and is illustrative of an individual farmer's care and attention, while industrial farms are large-scale and impersonal. She also argues that, while Carlson is right to suggest that it is possible to attune aesthetic appreciation to different appearances of the land, function in Carlson's sense is not all that should be taken into account in the appreciation of agricultural landscapes. An *aesthetic* response to landscape should not be disassociated from an *ethical* response. When considering the "functionality" of an agricultural landscape, we also need to consider what underpins it: is the agricultural landscape at which we are looking sustainable? Are those vast, rectangular sheds of different shades of grey concealing animal pain and distress? These questions about the relationship between ethics and aesthetics in agricultural landscapes apply equally, of course, to traditional as to intensive agricultural landscapes. After all, the stone walls that are seen as so full of character in Northern England are the legacy of the agricultural intensification of the Enclosures, with all the social problems that ensued. Brady's argument, then, draws aesthetic and ethical values together in thinking about the environmental ethics of agricultural intensification.

Pragmatist Environmental Ethics and Agriculture

Many of those in environmental ethics, in one way or another, espouse the view that elements of the natural world have non-instrumental, intrinsic value of some kind. But – with the exception of moderate concerns about animal welfare – these intrinsic values are deeply controversial, and are not widely held outside academic environmental ethics. Recently, a number of environmental philosophers have moved towards adopting a more "pragmatic" approach to environmental ethics. They take the view that while a discussion among philosophers about what is of intrinsic value may have its place within philosophical debate, in the world outside philosophy such discussions are likely to be theoretically unappealing and practically unworkable. Such pragmatists note that in debates about the future of agriculture outside academic environmental ethics it is *human interests* that are of central significance. An insistence on intrinsic value in nature, or even on the strong significance of aesthetic value in nature, is such a minority view that its dominance among environmental ethicists may lead to sidelining of the subject. Furthermore, Norton (1991) suggests, an appeal to intrinsic value in nature is not *necessary*. "The idea that other species have intrinsic value, that humans should be 'fair' to all other species, provides no operationally recognizable constraints on human behavior that are not already implicit in the generalized, cross-temporal obligation to protect a healthy, complex and autonomous system for the benefit of future generations of humans." That is, these values *converge* in actual cases. (This claim, though, as I will suggest shortly, is questionable.) Environmental pragmatists also suggest that these are good *strategic* reasons why arguments in environmental ethics should not be based on the intrinsic value of nature. Norton, for instance, maintains (in a North American context) that "a rhetoric that ignores human interest does not appeal to a broad enough slice of the American public to support a politically effective movement" (Norton 1991; see also Light and Katz 1996). The emphasis here is not so much on underlying philosophical principles but rather how to create successful value coalitions to work for politically effective ends.

This raises the possibility of approaching environmental ethics and agricultural intensification in a rather different way. The "intrinsic value" positions all began by identifying and expanding particular notions of intrinsic value, considered how such value might be "at stake" in agricultural intensification, and then worked out, based on these values, what should be done with respect to intensification. A pragmatic position is more likely to start with a particular issue (say, a proposed agricultural intensification project in a particular location) where a number of different value positions may be put forward in favor of, or opposed to, the intensification. Some of these value positions may advocate forms of environmental intrinsic value. Others, though, will have very different concerns, for instance concerns about the well-being of local people, or with distributional justice, or with the preservation of traditional farming methods, or with future generations of people. A pragmatic approach would adopt a wide-ranging pluralism potentially accommodating a variety of values, some of which the pragmatist may not personally endorse. Most environmental pragmatists are, actually, skeptical about the kinds of intrinsic values discussed here.[14] But in looking for a more practical level of agreement, there is no need either to adopt or to deny the underlying values held by others. What is being sought is a kind of *compatiblism* – where, in any particular case, different values can be seen as compatible in the contest of a specific environmental policy.

Such compatiblism is likely not only to be *value* pluralist; but also be *methodologically* pluralist – that is to say, the compatiblism will extend to include consequentialist, duty oriented and virtue approaches to ethics. Further the compatiblism is also

[14] They often prefer some form of anti-foundationalism, perhaps in the classical pragmatist tradition of Dewey.

likely to be pluralist with respect to *worldview*.[15] Quite different underlying world views may be able to converge at a more practical level. For instance, for some environmental philosophers the idea that aspects of the natural world carry intrinsic value forms an integral part of a broader worldview where seeing humans as fundamentally *managers* of the land is regarded as problematically arrogant. Others, though – perhaps working from within a Christian tradition of land stewardship – might affirm the idea that the natural world is intrinsically good as part of God's creation, but maintain that responsible management *is* the appropriate way in which humans should respond to intrinsic goodness. These worldviews seem at one level to be fundamentally opposed. Either humans should, or should not, be regarded as managers of the land. But, at another level, such different worldviews can be the foundation of a compatible *policy*. A principled policy of objecting to some particular form of intensification in a specific context, for instance, might be mutually agreed, even though for some this grows out of ideas about land value in the context of Christian stewardship, and for others from philosophical views opposed to arrogant human-centredness. Environmental policies, then, may be multi-determined, emerging out of a number of different worldviews, ethical approaches and values.

Several further issues also arise here. The first, and most obvious, is that such pragmatist positions tend to operate in a contextual manner. That is, decision-making about the ethics of agricultural intensification will reflect each particular, different context, considering the values that seem to arise out of and be "at stake" in the situation and are voiced by those involved. But this raises questions not only concerning the *nature* of the values expressed, but also *how* they are expressed. And further ethical questions about the processes by which some values emerge and are heard, the kinds of consultation and deliberation that take place, and the importance attributed to different voices (as well as how those who have no voices – for instance future people, children, animals – may be represented) suggest themselves. Second, it is likely that such value coalitions will be strategically organized around broadly-construed value terms that can be widely accepted by those adopting different ethical approaches and adhering to different underlying worldviews. One such broad value term, frequently arising in the context of agriculture, is *sustainability*.[16] I want to consider sustainability in a bit more detail before concluding, because to look more closely at sustainability may illustrate both the advantages associated with, and the problems flowing from, this more pragmatic approach to environmental ethics.

Sustainability, Pragmatism and Environmental Ethics

As Michael Redclift (1993) once argued, sustainability, like motherhood and apple pie, is difficult to disapprove of. It is hard to imagine anyone claiming to be "against sustainability". But sustainability is a very broad term, capable of a

[15] See Norton (1991) for a discussion of world news in the context of environmental philosophy.

[16] See, for instance, Cherfas (1996).

number of interpretation; and it is in this possibility of multiple interpretation that both its advantages and disadvantages lie. Most obviously, sustainability is usually taken to refer to the sustainability of human society and to imply at the very least some concern for the existence – and perhaps for the wellbeing – of future people. Cherfas (1996), for instance, maintains that "most agree that sustainable practices do not deny future generations their livelihood." The well known World Commission on Environment and Development definition (1987) characterizes sustainable development as development that "meets the needs of the present without compromising the ability of future generations to meet their own needs." Here, then, sustainability is seen as being about promoting and protecting the value of human lives, both present and future. But sustainability can also be construed to include concerns about sustaining particular human values – for instance, for the maintenance of traditional family farming or the protection of a particular kind of land aesthetic. And environmental concerns may also be incorporated into the idea of sustainability. The instrumental value of certain environmental states (for example of uncontaminated land or the maintenance of basic resources) for the sustaining of present and future human life may be emphasized. But *intrinsic* environmental values may also be included in the term "sustainability": environmental sustainability may be valued in its own right (for instance, in the context of the sustainability of valuable ecosystems).[17] Given the variety of values that sustainability may represent, it is clearly a term on which a number of different ethical positions could converge. It is likely to be compatible with different approaches to ethics, and can be underpinned by a variety of worldviews. And this compatiblism applies in particular as well as in general contexts. In any particular instance of agricultural intensification, sustainability could be the term on which a number of different values, ethical approaches and worldviews could converge: those adopting positions advocating intrinsic value in nature, those concerned with aesthetic value in the agricultural landscape, and those concerned about human well-being, in particular the well-being of future generations, for example.

But in this possibility also lies the *difficulty* of these kinds of value-coalitions. It seems, on reflection, likely that there will be cases where sustainability interpreted in one way *conflicts* with sustainability interpreted in other ways, or where Norton's (1993) claim about value convergence, noted earlier, does not hold. Suppose a particular form of agricultural intensification is proposed. If enacted, this intensification is likely to push one kind of endangered species further towards extinction. But on the other hand, it will assist the local struggling human population to provide for itself into the long term. The intensification seems to be unsustainable, seen from a species value point of view; but *not* intensifying seems to promote unsustainability from the point of view of the flourishing of a human community both in the present and the future. And it does not seem plausible that such conflicts between different values will never arise, even though they may arise less often than is widely thought to be the case.

---

[17] See, as an example, Scherer (2003).

The worry, then, about the pragmatist emphasis on value coalitions – or Norton's "convergence thesis" – is that on some occasions, at least, concerns about human values and concerns about environmental values issue in different environmental policies rather than converging on the same ones.[18] This may occur not just in the kinds of cases where human wellbeing seems to conflict with species values, but also where (for instance) there are proposals to intensify farming practices in ways that might have ecological benefits, but which change the traditional agricultural landscape in ways that some regard as destroying aesthetic value. Different environmental values may be in conflict with one another, just as human and environmental values may be in conflict; and adherence to one will issue in different policies than adherence to another. And, after all, this is not surprising, if one accepts that this is a world of plural values where a gain in some values may result in the loss of others.[19] This is not after all a "special" problem, relevant only to the environmental sphere.

Conclusion

In this paper, I have outlined a number of different values in, and approaches to, environmental ethics in the context of agricultural intensification. Much work in environmental ethics has focused on identifying intrinsic values in elements of the natural and living world. If environmental ethics is taken to have this focus then agricultural intensification may in some cases threaten, and in others protect or promote such intrinsic values. Responses to agricultural intensification from these perspectives will depend on which intrinsic values are being upheld and the particular details of any proposed intensification. Other environmental ethicists have challenged this approach, arguing for a more pragmatist, contextual understanding of environmental ethics, where policy can be decided on the basis of compatibility between different approaches. I have suggested that there are a number of cases where compatiblism will be problematic, though it is certainly a worthy aim. Indeed, there may be cases where those motivated by the desire to protect intrinsic values in nature would be more likely to achieve policy outcomes acceptable to them by the strategic adoption of human-centered values, even if those values are not their motivating force.

Agricultural intensification raises difficult ethical issues that inevitably go far beyond the scope of this paper. But even in this paper the contested and difficult nature of some of the values involved has become evident. Obvious as it may sound, there are no easy answers to the rights and wrongs of agricultural intensification from within any branch of environmental ethics – whether from those ethicists arguing for intrinsic values in nature, or for those adopting a more pragmatic orientation. However, it can be hoped, at least, that contributions from environmental ethics can help to clarify values at stake in and arguments about episodes and instances of agricultural intensification, and that some of the thinking in environmental ethics may contribute to decision-making in particular cases.

[18] This view is also put forward by Stenmark (2002).

[19] Here I'm adopting the kind of view put forward by Isaiah Berlin (1979).

Chapter 10
Agricultural Intensification and the Environment

Lawrence Busch*

Agricultural intensification poses a number of complex ethical issues. Put differently, one might say that producing maize ethically requires navigating through an ethical maze.

Comments on Professor Palmer's Chapter

Clare Palmer poses an excellent overview of the issues in (agro) environmental ethics. She notes that environmental philosophers have devoted little or no attention to agriculture. (Thompson's work [1995, 2007] would be a significant exception.) But the same appears to be true of ecologists. Some 30 years ago Mayer and Mayer (1974) wrote of "Agriculture: The Island Empire," decrying the separation of agricultural issues from the mainstream of modern science and especially from ecology. Today the situation is somewhat better, but a considerable chasm remains (Vandermeer 2003). Most ecologists have a passion for tropical rain forests, but steadfastly avoid cultivated fields, whether in the tropics or elsewhere. Similarly, anthropologists may be said to have a fascination with kinship systems, but – despite studying agrarian societies – rarely study agricultural practices of any culture.

In some ways agricultural scientists have fared somewhat better. Most agricultural scientists, overproduction notwithstanding, are still interested in increasing the production, productivity, and efficiency of farming. But things are changing at the edges. On the one hand, the advent of molecular biology has transformed crop and animal science, creating new links between production-oriented scientists and their colleagues in other fields of biology. On the other hand, the growing interest in sustainable agriculture, organic farming, rotational grazing, and in measuring the environmental consequences of genetic modification have enhanced interest in ecological research and methods (Robertson et al. 2004).

---

* University Distinguished Professor in the Department of Sociology at Michigan State University and Professor at the Center for Economic and Social Aspects of Genomics at Lancaster University.

P.B. Thompson (ed.) *The Ethics of Intensification*,

Yet, in the agricultural sciences as well, sustainability has become something of a buzzword, usefully trotted out whenever environmentalists are within earshot. Full integration of ecological reasoning into agriculture remains more a goal than a reality.

Under these circumstances, it is no surprise that environmental philosophers have not paid much attention to agriculture. Yet, agriculture is (among other things) but one form of environmental management. Others include forestry and fisheries, but also more passive means of managing the environment such as the construction of hydroelectric power stations, vast irrigation and urban water supply systems, through generation of chemical particulates in the air and water, usually referred to by the morally charged term: pollution.

Agroecosystems

Professor Palmer discusses the problems associated with valuing of ecosystems, whether of the agricultural sort or otherwise. But this masks what I would argue is a more fundamental philosophical and practical problem: How do we know when we have an ecosystem? Individual organisms are generally easy to identify and have more or less clear boundaries that mark them off from the rest of the world. Species of higher organisms are particularly easy to demarcate. I have never seen anyone confuse an elephant with a giraffe. But single celled organisms tend to have rather blurred species lines. Biologists can and do make or break entire careers attempting to produce clear boundaries.[1] (Agro) ecosystems are even more problematic (Bawden 1990; Bawden and Ison 1992). Do they exist as distinct entities or are they merely heuristic devices? If they do exist, where do they begin and end? Should they include only the crops and animals of interest to humans? Or should they include the various insects, diseases, rodents, bacteria, and other organisms that inhabit cultivated fields and domestic animal housing? What criteria should be used to define them and distinguish among types of agroecosystems? Is an intensive agroecosystem one that includes many cultivated plants? Or is intensity better defined as including all the organisms and the geophysical characteristics of an agricultural field? These are questions that demand answers both from philosophers and from practitioners who wish to understand and maintain a given environment.

Biodiversity

Professor Palmer also suggests that agricultural intensification could have a deleterious effect on biodiversity. Indeed, there is substantial evidence to suggest that large monocultures covering thousands of hectares reduce the diversity of plants, insects, birds, and microbes. However, it is worth noting that the reverse is true as well. That

[1] For a history of some of the many debates about species boundaries, see Hull (1988).

is to say, as human population increases require the production of more and more food, it is quite likely that *both* intensification (i.e., more intensive use of existing farmland) and extensification (i.e., greater use of land previously excluded from agriculture) will occur simultaneously. Extensification comes with its own problems. First, new lands brought into cultivation are likely to be those least suitable to agriculture. They will have more erodable soils, with fewer nutrients, with thin layers of topsoil. Second, all other things being equal, these new lands are likely to be home to more species than lands subjected to agricultural production for centuries.

Animal Welfare

In contrast to the genetic modification of plants to permit more intensive production, genetic modification of animals raises yet another class of questions. Unlike plants, animals are sentient beings. Domesticated animals are both a source of food and of human responsibility in several ways. First, we must care for domesticated animals in order for them to survive. Caring for them in ways that create needless suffering (both in raising and slaughtering animals) quite obviously ignores the welfare of those animals. But at the same time, it also says something about us as human beings. In brief, those who subject animals to needless suffering are violating some closely held and widely shared values. Certain forms of confinement feeding of animals appear to pose these problems.

Second, intensification of animal production poses another related set of problems. As both Sandøe (Gamborg and Sandøe 2003; Sandøe et al., 1999) and Gussow (1991) have noted in different ways, animals have a certain integrity that comes with sentience. For example, a minor modification in the genetics of animals (e.g., decreased fat content, more edible parts per unit of weight) may be acceptable as it remains within the range of 'normal' variation of the animals. In particular, it does not usually involve the introduction of genes from other species. However, radical modification of animal genetics (e.g., featherless chickens) appears to violate the integrity of the organism. The boundary between the normal and the unacceptable here is, of course, blurred. One may use the techniques of genetic modification to produce phenotypical modifications that remain within the 'normal' range or to make far more radical changes. And, that is in itself a part of the problem. Some would argue that genetic modification leads one down a slippery slope toward treating animals as mere objects to be manipulated as we wish to fulfill human desires. Pushing the argument further, we see that such behavior might easily lead to treating human beings in the same fashion.

Aesthetics

Another aspect of intensification is aesthetics. Some landscapes – including agricultural landscapes – are seen as aesthetically pleasing while others are seen as ugly. Oftentimes, the aesthetic and the ethical, i.e., that which is beautiful with that which is good, are conflated. Thoreau (1854 [1995]) argued against railways which he saw as ugly intrusions into the serenity of the countryside. Yet, today railways are often looked at in a nostalgic manner, and pictures of trains crossing the prairies conjure up romantic images. We would also do well to remember that in the 1930s, landscapes with smokestacks were viewed quite positively; only after the environmental movement began did smokestacks begin to appear as ugly blots on the landscape. And, today pictures of terraced rice fields in Indonesia or China – some of the most intensively cultivated places on earth, or of manicured Swiss cow pastures, are often hung on office walls affirming their beauty, and by implication, affirming (rightly or wrongly) their ethical value.

In a now classic study of several years ago, Dutch sociologist Jan Douwe van der Ploeg (1990) noted that in some places divergent aesthetics and ethics can be found. The highly productive and wealthy Emilia Romagna valley in northern Italy is one such place, where both intensive and extensive agricultures can be found side by side. As Ploeg notes, in most studies intensive agriculture has been equated with better agriculture. Yet, from the vantage point of extensive farmers in the Emilia Romagna valley, such agriculture is plagued with technical problems. Conversely, those who practice intensive agriculture there emphasize craftsmanship. Ploeg is silent on the environmental consequences of either type of production, yet there is little doubt that the two forms of agriculture produce different environments.

From Supply to Demand

One aspect of agricultural intensification not discussed at length by any of the speakers is the dramatic shift that is occurring worldwide from a supply-driven agricultural economy to one that is demand-driven (Busch and Bain 2004; Nicolas and Valceschini 1995). Until recently, farmers, ranchers, and fishers not only in so-called developing nations, but in highly industrialized nations as well, could and did produce whatever their forebears had produced. While this did not necessarily lead to high income, it was possible to get by simply by supplying the market with whatever one could produce.

Moreover, most markets were spot markets – whether the open-air wholesale markets still common in poor nations, or the huge warehouse-like markets in which goods of various qualities were stacked neatly in rows of cartons. Prices depended largely on the volume demanded by buyers and the size of that season's harvest. Retailers, in turn, sorted by hand through large volumes of product in an attempt to get the best quality – however defined – for the lowest price.

However, this world of supply-driven agriculture is fast disappearing. It is being replaced by a largely new demand-driven agriculture.[2] Let me briefly outline the

[2] Of course, suppliers to the wealthy have always relied to a great degree on demand to drive those markets.

nature of this phenomenon. The first and still continuing change was a rapid concentration in the supermarket and fast food sectors. Consider the case of supermarkets. In the US, the top five chains now control at least 40% of the market. In Britain four chains have 70% of the market. In Australia, three chains have 70% of the market. Nor is this economic concentration limited to industrialized nations. In Chile, the top four firms have a 66% market share. Kenya has two supermarket chains with more than 20 stores each. And, even Zambia, a poor country with few inhabitants, has a chain with 18 stores. Globally, supermarkets are serving a greater segment of the population, and a small number of chains dominates the marketplace (see, for example, Reardon and Berdegue 2002).

Furthermore, as it became clear that tariffs and quotas on food products would decline, the larger food retailers – supermarkets, fast food restaurants, and caterers – began to invest heavily outside their domestic market areas. Soon, as a result of the forming of the World Trade Organization, the largest retailers were operating in several or even several dozen nations at once. Three giants emerged globally:

- US-based Wal-Mart has about 1,400 stores, mostly of the hypermarket format. It is the largest chain in both Mexico and Germany and is third largest in the United Kingdom.
- Dutch-based Royal Ahold has about 5,600 stores on four continents, ranging from small convenience stores to hypermarkets.
- French-based Carrefour has about 9,000 stores in 30 nations, mostly using the hypermarket format.

These three giants soon began to source and to supply stores globally. In so doing they discovered that agricultural supply chains were, from their perspective, extremely inefficient. Goods were handled many times resulting in high rates of spoilage. Product standards (for product quality, package sizes, food safety, phytosanitary requirements, etc.) varied from one country to the next, resulting in rigidities in the movement of agricultural goods. Numerous brokers each took their commissions on goods bought and sold. Supply was often poorly articulated with demand resulting in wild price and quality fluctuations. They soon realized that most of these hurdles could only be overcome by radically restructuring the entire supply chains for agrifood products.

Thus, major changes have occurred and continue to occur including:

1. Sourcing directly from farmers, thereby eliminating scores of brokers, shippers, and the like
2. Requiring suppliers to deliver a certain quantity meeting standardized specifications at a time and place desired, thereby creating a new form of intensification
3. Stocking products on store shelves all year by sourcing counterseasonal products for US and EU markets in the southern hemisphere, thereby intensifying production in regions where previously production was far less intensive
4. Increasing the variety of products on store shelves, in an effort to attract more customers via non-price competition
5. Establishment of store brands which incorporate particular varieties, cosmetic qualities, and private standards (about which more below)

Furthermore, although competition has remained fierce and profit margins low, like good oligopolists in other industries, these firms began to compete on more than price alone. Various aspects of quality, including easily observed aspects such as cosmetic quality, quality attributes learned through experience (e.g., taste) and what economists have called credence attributes (e.g., food safety, fair treatment of labor, etc.) have become grounds for competition.

The fast food and catering industries have followed a similar trajectory. McDonald's, for example, specifies the potato varieties for its french fries. Thus, the shift to a demand-driven agriculture has been accompanied by a grafting of what has been called an economy of qualities on top of the existing economy of quantities (Eymard-Duvernay 1995). More and more, especially in the areas of fresh meats, produce, dairy products, and fish, quality differences abound.

Standards and Intensification

As concentration in the retail sector has continued apace, retailers have begun to develop their own standards for food safety, food quality, environment, labor, animal welfare, organic, and fair trade, to name but a few. Standards are not merely technical devices that may be used to improve the flow of trade by reducing transactions costs. They are also strategic means of restructuring markets to achieve a variety of ends (Busch 2000; Reardon et al. 2001). Thus, standards can profoundly affect the degree to which agriculture becomes intensified as well as the paths taken to intensification.

Retailers have been motivated in a variety of ways to create and enforce private standards. First, various private voluntary organizations (PVOs), including consumer and environmental groups, have pressured retailers to adopt or reject certain practices. Given the low profit margins in the retail food industry, fear of losing a small percentage of one's customers can be a powerful reason for change. Second, as retailers become larger, protection of one's brand names becomes more and more important. Practices deemed unacceptable by even a small portion of consumers can result in a considerable decline in sales. Third, large firms are easy targets for many lawsuits. Insisting that suppliers meet certain common standards can and often is a good defense against legal action. Fourth, many nations (e.g., Great Britain) now have 'due diligence' clauses in their law. Such clauses make it necessary for a retailer to show that they have done 'due diligence' in order to escape unscathed from a plaintiff's claims. Standards for suppliers are one important means of showing due diligence. Fifth, retailers desire to position themselves as protectors of consumers and the environment. Requiring suppliers to meet standards and conveying that action to consumers is a means of demonstrating their good faith in meeting consumers' concerns. Finally, as implied above, standards allow retailers to differentiate their products from those of competitors. Requiring particular crop varieties, methods of production and processing, or particular attention to health and safety of farm workers, can help retailers to maintain a competitive edge in the highly competitive global food markets.

Third Party Certification and Intensification

The growing plethora of standards might well have highly desirable effects on the environment, but it may equally easily create a new set of environmental dilemmas. A great deal will depend on the means by which standards are developed, monitored and enforced. Retailers appear to be leaning toward what is coming to be known in the industry as Third Party Certification (TPC) (Tanner 2000). Specifically, retailers are reluctant to play the role of policemen, especially with the very same persons with whom they are engaged in market relationships. They would much prefer to let others do the policing, since this also tends to shift liability to others. Thus, over the last decade a plethora of third party certifiers have arisen to meet retailers demands. Certifying firms include long-established enterprises such as NSF International and the Société Générale de Surveillance (SGS), as well as small start-up firms eager to capitalize on the minor boom in such certifications. Moreover, some retailers demand that suppliers use particular certifiers, while others allow suppliers to choose. Some have detailed requirements, while others appear quite casual about certification requirements.

One thing is certain. Retailers' desire to employ and insist that their suppliers meet standards for an increasing array of product and process characteristics is transforming the agricultural sector rapidly. Producers and researchers in developing nations are likely to face particularly high hurdles as they attempt to meet ever higher standards (Busch 2002). The forms that agricultural intensification takes as well as the ethical problems that it poses are likely to be shaped by the move towards certification. Without a doubt, ethical issues encompassing distributive justice, environmental justice, the rights of farmers and farmworkers, animal welfare, and biodiversity of crops and animals, among others, will emerge. Now is the time to raise the tough questions, before particular policies and actions are cemented firmly in place.

Chapter 11
Agricultural Intensification, Environmental Ethics and Sustainability: Some Ethical Observations

Nigel Dower*

This paper arises out of comments which I was asked to make on Clare Palmer's conference paper on environmental ethics and out of further reflection on the theme of the conference.

General Observations on Sustainability as a Dimension of Environmental Ethics

Environmental Ethics: Broad and Narrow Accounts

Clare Palmer offers a detailed, clear and balanced account of the issues which are raised for environmental ethics by agricultural intensification, and I am in broad agreement with most of the many significant points she makes. I only take issue with her on a number of small points of detailed exposition which I discuss in the second half of the paper, and on the general way she sets up the discussion in terms of her understanding of environmental ethics which I take up first.

For the purposes of her paper she takes environmental ethics to be about what is considered to be of intrinsic value in the environment, and discusses the impacts, positive or negative, on four cases of carriers of value in the environmental – individual sentient things, individual living things (respect for life); species of living things and whole ecosystems. Palmer's paper focuses then on environmental ethics as about the intrinsic value of things in nature and the impact of agricultural intensification. But given the general title of her paper included 'environmental ethics', I feel that before I turn to her particular points in the second half of this paper, I should like to look at the issue of environmental ethics more broadly as concerning the impact of human activity on the environment from the point of view of the well-being of human beings.

---

* Honorary Senior Lecturer in Philosophy, University of Aberdeen

P.B. Thompson (ed.) *The Ethics of Intensification*,

In this regard much of the interest in environmental ethics links it to the idea of sustainability or sustainable development. It is commonly assumed that what is to be sustained is development itself and the conditions of human-well-being. This assumption however turns out to be unduly narrow, as is revealed if we ask questions like: What form of agriculture is sustainable? Can intensification contribute to making agriculture sustainable?[1] Whilst Palmer turns to the issue of sustainability towards the end of her paper, I prefer to look at it first. The issues she raises are unlikely to be taken seriously by policy-makers unless they are situated in the discourse of sustainability, and can be framed by the question: what values are to be sustained in development?

What Is Sustainable Agriculture?

The question whether agricultural intensification constitutes a form of sustainable agriculture turns on what it is we wish agriculture to sustain. It is no doubt obvious but also obviously inadequate to say that it is *agriculture itself* that we want to sustain. There are aspects of agriculture as a primary practice (done by farmers on a farm) that are crucial to the sustaining (or non-sustaining if they are in fact not appropriate) of that practice, like methods of cultivation and harvesting or types of resources like animal feed, seeds or fertilizer/chemical inputs used. But that is really only the tip of the iceberg.

First, the kind of agriculture we want to sustain has to be seen as appropriate to the wider goals it serves – food for growing populations, economic growth or development as freedom, for example – and so what we need to sustain is much more than what is itself capable of being sustained into the future. Thus at its simplest, low yield agriculture might be – and certainly was in the past in many areas – ecologically sustainable, but simply did not produce enough for human needs. So the challenge is to increase production whilst remaining ecologically sustainable (or returning to a sustainable state, if we are not in one now). Much of the rationale for and justification of past intensification has been precisely because of the need to serve expanding human needs.

Second, agricultural practices, including intensification, occur in a vast matrix of causes and effects, in which various things need to be supplied as inputs and in which various outputs occur such as the sale of agricultural produce in both domestic and international markets. Since agricultural activities both create demand for various inputs and in turn affect what others do with their outputs, the ethical

[1] There is another question which might be asked: is agricultural intensification itself sustainable? This is different question since it is about the process of intensification being itself sustainable. One rather supposes that beyond the near future, it might well be that the process could not go on further since there are probably limits. However the *effects* of new forms of intensified agriculture may go on indefinitely and be sustainable – and that is what we are more interested in.

assessment of agriculture is partly about these wider factors, which might be both environmental and social, and also both local and global.

We should note here that, whilst it is undoubtedly true, as Thompson well documents in his paper, that in the past intensification has played a crucial role in the expansion of human populations and the increase in general human wellbeing, we cannot extrapolate from this and say that continued intensification will continue to reap benefits. It is precisely our modern situation which creates the problematique. Environmental limits may have kicked in now in a way never before experienced. Concern for non-human welfare may raise questions which were not asked before. Modern highly mechanized agriculture may have unequal effects on human well-being so as to raise issues of social justice rarely faced before. The genetic modification of organisms raises new sets of ethical issues. So one may wish to adopt a cautious approach and judge intensification on a type-of-case by type-of-case basis, taking a middle road between technological optimism and technological pessimism.

Furthermore, in assessing the varieties of intensification we need to recognize both the local and global dimensions involved. Much assessment will focus on local conditions and circumstances; but as I have already noted, the assessment of intensification depends heavily on the wider network of consequences including global consequences.

Parameters of Sustainability

For any proposed form of agricultural intensification, we need then to consider at least three parameters:

What are the inputs?
What are the salient features of the agricultural process itself?
What are the outputs, both intended and unintended?

By the inputs I mean things like: plants, seeds and animals, including genetically modified forms of these; fertilizers, pesticides, insecticides etc.; machinery of various kinds; skilled labor. Ethical issues can arise especially in regard to the biotic inputs, both because of the issues to do with GMOs per se, but also to do with the control over the development of seeds etc., as evidenced particularly in the use of patenting and the development of TRIPS conventions in which there is often an undervaluing of the communal knowledge about indigenous seeds built up over generations of practice. The development of seeds etc. by large companies raises wider questions about *what* seeds are developed, given that the needs of poor people in the South may not be the same as those of the North.

By the agricultural process, I mean what actually happens on or near the farm and its effects on what is on the farm. What happens to the land or its soils may be of ethical concern e.g. in regard to applications of inputs such a fertilizers and pesticides. What happens to the biota, particularly animal biota, are now of great concern in the animal welfare lobby. Human labor is employed in various ways, and this raises interesting questions about freedom and exploitation in relation to large-scale operations in the third world. Furthermore the way the farmland is used, what parts of it left fallow or wild, may have impacts of ethical significance on the wild biota living on or near the farm.

By the outputs, I mean the production of food, timber and other products which go to local markets, national markets and further afield. Various ethical issues are raised here about the distribution of food, food subsidies and the effects on international trade, food dumping etc. on the one hand, and about the patterns of affluence sustained by such modern food production systems which for many environmentalists raises serious problems – e.g. the assumption that food should cross the world to satisfy our pallets, that the availability of endless varieties of food fosters unsustainable lifestyles (if generalized and seen in the longer term).

The last claim about sustainable life-styles points to a more general environmental limit which is now being increasingly recognized, namely climate change. Agricultural intensification may or may not contribute to higher levels of carbon-based fuel use overall. If an intensification uses more fuel but also produces more food for more people, its overall impact could be neutral or even positive. But the real problem lies elsewhere. Insofar as agricultural intensification is part of an overall network of inputs and outputs in which more things travel greater distances and/or more things are needed for materially more affluent life-styles, then there is a serious sustainability issue at stake here.

Environmental Sustainability

Many of the above issues relating to the three parameters have to do with environmental concerns. But within the term 'environmental' there are several different perspectives. Many of these relate to broadly anthropocentric concerns about needing to sustain the environmental base for human well-being, whether for one's own future, for future generations or for the wider society of all human beings living now (given that many concerns about agriculture and development more generally are often more focused in practice on one's own community or one's state). But equally many of the concerns can also be construed as based on biocentric considerations – on the impact of agricultural practice on higher animals (whose well being is seen as intrinsically valuable), on other forms of biota and on land or soil seen as an ecosystem valuable its own right; and on the attitudes human beings should have toward the land they work, which should be and often in the past was one of being part of a community, rather than the land being seen as often now as a resource for maximizing profit (this issue of attitudes being linked to Thompson's conception of virtue ethics).

One way of expressing the points in the previous paragraph is to ask: what are the values to be sustained in sustainable agriculture?[2] Are they various kinds of human-based

---

[2] Sustainability presupposes other values. Its value is parasitic upon the value of the things to be sustained.

values or do they include other kinds of values which an acceptable form of agriculture or agricultural intensification must take into account? Whether all these values are to be seen as the province of *environmental* ethics or whether only the second domain of biocentric values is its proper domain is a matter for terminological debate. I am inclined to regard all these concerns as environmental ethics, whereas I sense that Palmer wishes to restrict the domain. At any rate her concerns are primarily to do with the various ways non-human values come into the ethical reckoning and how this impacts on agricultural intensification. To this I now turn. She presents a lot of very nuanced considerations and arguments, and I can only touch on some of these in what follows.

Comments on Palmer's Paper

Sentient Life

First I want to address the question which she raises: is it better to have fewer intensively raised animals involving very painful lives or more less intensively raised animals living less painful lives (though not good lives)? This question is misleading if it implies that one way of proceeding would, if better, be right. Quite independently of stronger vegetarian arguments for not killing animals as such, it is surely wrong to bring into existence animals whose lives overall go badly. This is why the reply to vegetarians that if everyone were vegetarians, many animals simply would not exist misses the point, if the animals in question have, as often they do in intensive farming regimes, on balance a painful and stressful existence.

If the only two alternatives are bringing into existence animals with expected very painful lives and animals with predominantly painful lives, then neither route ought to be taken. This is not because such animals are *harmed* by being brought into existence, since it makes no sense to say of any being that bringing it into existence either harms or benefits it since there is no relevant comparison in being non-existent. Nevertheless if by bringing into existence a being which we intend (either by our controlling its life once born or by our having predetermined it to have a certain kind of life by prior genetic manipulation) to have a painful and stressful existence, then it *experiences harm*. Furthermore if – though I doubt if this conclusion is conceptually necessary – harms have to be caused by something that 'harms' the subject that is harmed, these are caused by the biological and environmental conditions they are in, if not by human beings *who could have acted otherwise*.[3]

Here as elsewhere we see the importance of recognizing both the deontological and the consequentialist aspects of the evaluation. If it is simply wrong (deontologically) to bring into existence beings whose lives will on balance not be worth living (because they are filled with stress and pain), then we should not proceed with the

[3] In this sense, to take issue with Parfit's famous argument, we do harm future generations even though other policies would not have brought them into existence (see Parfit 1991).

agricultural method involved, either as it is or intensified. If on the other hand, without taking the Regan line that the raising of animals as such is unjustified control over them and so wrong (Regan 1984), we recognize that new developments in agriculture mean that fewer animals would lead lives which, whilst containing some pain and stress, were on balance reasonably pleasant lives compared with more animals in the same conditions, then it would be strange not to regard it is right to prefer the course that involves less overall pain and stress in the world.

This is relevant to another point which Palmer makes about value conflicts in respect to wild animal welfare and other environmental values. If agricultural intensification (and this is a big 'if' since the alternative might not be returning areas of land to relatively unmanaged status) led to more flourishing of wildlife in neighboring areas, would this justify the worse condition for farm animals? If my argument above is right, the answer is 'yes and no'; yes, if what is proposed for farm animals is still a life worth living, but no, if what is proposed means lives that should not be brought into existence in the first place – which may be an argument either merely against intensification of what is happening or more generally against the existing farms practices as well. However it should not be supposed that the reason for this is that wild animals have greater value because they are wild – an approach sometimes favored by those who see a big gap between concern for the environment and concern for animal welfare. The issue concerning wild-ness is taken up again later in the discussion.

The question which Palmer asks raises a more general point about moral responsibility which she assumes rightly both here and at other points in her discussion. We are responsible not only for the effects which we intend or aim at, but also for the likely effects of our actions which are not the objects of out intentions. For example we are responsible for the well-being of those animals whom we both control and intend to be produced for food etc. as well for the welfare of those animals (and other forms of life) which are significantly affected by our actions though we may not 'control' or 'manage' them. This actually applies both to what happens *within* the geographical space – here the farm – as well as to what happens *outside* that space, for example in the neighbouring areas adjacent to the farm – but not only this of course, since our environmental footprints may be very extensive, even global.

Does biotechnology alter any of this? Many developments in biotechnology have of course led to breeds of animals that have rather bad lives and this is ethically highly problematic. What if biotechnology could or can create new breeds of animals that were happy with their restrictive living conditions or were even insensitive animals? Clearly many animals around us have, through generations of breeding, developed different natures or 'teloi'. A dog now would probably not have a 'good' life devoid of human company whereas his wild ancestor would have shunned human company. Cows would not survive or would have less satisfactory forms of existence if turned into the wild since they have been bred to have a symbiotic relationship with humans and certain forms of human intervention. So if biotechnology contributes towards modifying forms of sentient animals which are capable of a *good* life in symbiosis with human beings and subject to human control, then, so far as this argument goes, that is ethically satisfactory.

But it may be that the question 'what is wrong with creating contented chickens in cages or insensitive beasts?' is the wrong question to ask here. If we have accepted that the control of animals is legitimate for the production of food etc., then maybe we have to say 'these developments are to be welcomed', but maybe other issues are raised. Perhaps there is an aesthetic reaction here: a happy caged chicken offends our image of a healthy chicken as does the idea of an insensitive beast our idea of an 'animal'? Calling it aesthetic is not to dismiss it as unimportant, but perhaps we need to say rather more than this. It seems to represent an attitude of sheer control or manipulation of nature (as biotechnology is seen by many to be quite generally) and thus to be a failure of moral virtue, even if it is not straightforwardly wrong. But there is a further consideration, which leads us into the second broad category that Palmer considers, namely biocentrism and 'respect for life' as such.

One thing that might disturb us about these examples is that the lives created seem somehow less rich than they would otherwise be: a chicken happily sitting in its cage has a less full life than one running around the yard, we may say; an insensitive beast, whilst alive in a complex way, has no life at all (in Regan's sense of being a subject of a life). We make similar judgments about human lives: someone, whether through genetic make-up, brain damage or limited early conditioning, who has very few desires may seem to lead a less full human life even if these desires are satisfied than someone who has a wide range of developed powers and capacities. I think we all feel the force of Mill's contrast between the fool and Socrates, even if we distance ourselves from the elitism it probably expressed. Not all human lives are equally rich, and whilst humans have a remarkable capacity for compensating if certain powers are missing or lost (like hearing or sight), there are significant differences in how qualitatively full various lives are (whilst we accept morally the equality of all). Now one may ask whether it is morally all right to deliberately bring into existence beings whose lives have less richness then they (or their substitutes) might have had. This leads us into the 'respect for life' approach because it is not obvious that if one accepts that all life has value, one needs to accept that all life has *equal* value. A bacterium may fulfill its telos admirably, as may a dandelion as much as an insensitive beast, but would the realization of these teloi have the same status as that of a human, a lion, a cow, or even a bee?

Life as Such

The only point I want to make about the impact of intensification on life in general is to say that I am not as sure as Palmer that concern for life here is primarily deontological. Certainly the general claim that all life has intrinsic value or is morally considerable does not entail that that value provides an absolute side-constraint on what I ought to do, or even a side-constraint that is sufficiently powerful that I can only interfere with the living thing or destroy it *if* I have *very* powerful reasons such as the need for survival. This covers self-protection against attack e.g. by lions or microbes, and self-preservation including centrally the need for food, not just for me now but for human beings generally and into the future.

It may be that we should accept a weaker deontological claim that we must never destroy or damage any form of life without proper justification, that is we should not act *carelessly* and that we should have the attitude that whatever we do, like weeding the flower bed or walking along a path, has impacts on life. However it seems to me that there enters here a form of consequentialist reasoning that we are trying, if we have a respectful attitude to life, to minimise our impact negatively on life (by not destroying life unnecessarily). Taylor's 'minimum harm' conception points to a form of consequentialism. One may say that 'if there are lots of wrong things we cannot avoid doing, let's do less of them' is still deontological but in fact it is a kind of minimizing 'moral bads', and one could just as easily say more straightforwardly that one should not do what destroys or damages what has intrinsic value less than what one might do. But this form of 'negative utilitarianism' is really only a small distance away from the recognition that we also have duties positively to promote intrinsic value. This need not be the problematic 'maximizing' conception of utilitarianism but merely the recognition that sometimes we have duties to promote biotic value.

Species

Turning to the value of species as such, it is clear that, whether we value species as intrinsically valuable lifelines (rather than as classes of similar things) or we regard them as instrumentally valuable to human beings present and future, agricultural intensification is not the main factor which is leading to the reduction of natural species. Rather it is agricultural extensification, e.g. in the clearing of areas of rainforest for farming often for traditional low intensive farming and more general 'development' (land for housing etc.). But as Palmer indicates, intensive farming could have good or bad effects on the protection of species. She raises an important question about the relative value of natural species and those which might be created by genetic engineering. If species have intrinsic value, would the artificially created species have intrinsic value and if so would it have the same amount as naturally occurring species? Of course insofar as new breeds of plants and animals have developed over the centuries, the same question can be asked of domesticated breeds versus those occurring in the wild.

There are really two questions here. How important is the history of something to its current evaluation? If history is relevant, why is the distinction between a history in which humans play no part and a history in which they do play a part relevant? One can think of cases where the history of something is important, for example a gift that was stolen by the donor or consumer goods that are produced by sweated labor etc., but why is the distinction between natural and artificial – artificial in the basic sense of being produced or modified by human artifice – so relevant? Perhaps it is relevant to our aesthetic appreciation. For some the wild-ness of a species or landscape is part of what makes it attractive, as much as the *non*-natural-ness of painting – its being caused by certain intentions and exercises of skills – makes painting attractive to others or even the same people.

But equally the aesthetic appreciation of either or both may be much less dependent on such beliefs about their history for many others. But these aesthetic values seem to be rather different from the assessment of the intrinsic value of a species which, if it exists, does not seem dependent on its long-term history – though, if Rolston is right, a life-line must logically of course have temporal depth to be a life-life at all. Palmer later considers aesthetic appreciation of nature and similarly accepts that this is a separate category of evaluation from the recognition of intrinsic value. Whilst others might disagree partly because they might dispute the implied contrast between subjective and objective value, this distinction seems to be right, though I would add one caveat. Leopold's idea of beauty as one of the three value features of the biotic community was probably seen as an objective feature which, whilst the appropriate object of the value of aesthetic appreciation, was something more than that and an intrinsic value.

Ecosystems

In regard to ecosystems it is clear, as Palmer argues, that agricultural intensification could contribute to sustaining or restoring ecosystemic value and that it could do so in two senses: first, insofar as such intensification is consistent with or conductive to areas of (relatively) wild or unmanaged areas adjacent to farms remaining healthy, balanced and flourishing or becoming more so; and second, insofar as certain kinds of practices on the farms themselves may contribute to the values of ecosystems because the farm is part of the ecosystem. On the other hand one must also recognize that all too often intensification leads to the opposite results. She quotes Rolston as recognizing that an agricultural area could be 'managed sustainably' and thus could have an 'ecological fit' with the surrounding ecosystems. For Rolston I think it is concession that the farm itself, whilst not part of the ecosystem, could be as it were compatible with (rather than damaging) surrounding ecosystems. By contrast for me (and for Palmer I think) the important point is that farms are part of ecosystems and insofar as farm practices are dominated by mechanization, huge use of chemicals and so on, they make the ecosystems less healthy or flourishing, and it is important to see human beings not as intruders but as part of the ecosystems *including* those parts that they manage.

The Issue on Convergence

More generally it is important to recognize, especially in regard to concerns about species and ecosystems, that whilst the discussions here, following Palmer's concerns, are about the effects on these regarded as having intrinsic value, the concerns that many environmentalists have about these effects are not centered on this kind of consideration, but on the effects on human well being, now and in the future. It seems to me that all the worries she has raised can be raised by enlightened anthropocentrists as well, especially those who have regard for the issues of sustainability. This links up with Palmer's conclusion that in many areas there may be convergence of what should be done in respect to the environment. The convergence may not merely be between environmental ethics, development ethics and aesthetics, but also *within* environmental ethics itself, which, as I suggested earlier, is driven by human concerns as well as concerns for non-human values. Whilst Palmer is right that there will not always be convergence and that focusing on sustainability does not avoid this divergence since different and incompatible things may be pursued under the banner of sustainability, the acknowledgement of convergence or compatibility is important, partly for pragmatic reasons about getting policy makers to act. But it is also partly, I would add, for theoretical reasons to do with the conceptual conditions for accepting a 'common' (though not necessarily *universally* accepted) ethic which may have a number of different methodologies or worldviews supporting it (see Dower 2005). It is convergence on shared values and norms that is significant, not just the more accidental or contingent sharing of policies or decisions which the word 'compatibility' suggests.

It is important to stress this because many people interested in agricultural intensification may dismiss the concerns discussed in Palmer's paper and mine as 'all very well for idealist, but completely irrelevant to real policy issues'. But if there is widespread convergence on the need to relate to the environment in certain ways that challenge prevailing assumptions, that is a different matter. Indeed, apart from the theoretical relevance of the issues Palmer raises and the importance of this perspective being included in any publication on agricultural intensification, it is important to recognize that there are a significant minority of people for whom these biocentric considerations have weight and that their preferences which are informed by these value judgments do enter the public forum, explicitly and implicitly, when policies get shaped.

But the point about compatibility or convergence is really about two phenomena, which may be called 'over-determination' (a term Palmer used in an earlier conference version of her paper) and 'multi-determination'. Over-determination is the fact that for any given thinker she might have environmental reasons, developmental reasons and aesthetic reasons for a policy where each type of reason on its own would have been sufficient to shape her judgment or motivation. Multi-determination is the fact that each of these reasons (and many others) are acceptable to *different* thinkers who may then agree on what is to be done and indeed on a range of core values *but for different reasons*. The desirability of the latter reflects a much wider interest in generally trying to find shared values on many issues which can be supported from many different worldviews or starting points. The interest in trying to find a 'global ethic' is another expression of this ethical imperative of our time.

Chapter 12
Animal Welfare and the Intensification of Animal Production

David Fraser*

Introduction

The intensification of agricultural production has been a focus of debate because of its profound effects on food availability, rural populations, use of resources, biodiversity, and many other issues. In the case of animal production, however, the debate over intensification has taken on a distinctive tone because at the centre of the process are animals. In many cultures animals are viewed, at least to some extent, as sentient beings with interests of their own. Moreover, animals are the focus of ancient moral beliefs about the relation of humans to the natural world and about proper human conduct towards other beings. Thus, to confront ethical concerns about the intensification of animal production, we need to understand how intensification affects animals and their welfare, and how it relates to ethical beliefs about the care and use of animals.

In this essay I will explore different ideas about the welfare of animals and the roots of these ideas, examine some key features of the intensification of animal production, look at some traditional ethical ideas about animal care in order to help explain why the intensification of animal production has become such a prominent social and ethical issue, argue that some of the standard claims made by critics of intensive animal production are flawed, propose an alternative hypothesis to account for key developments in the intensification of animal production, and identify some actions to improve animal welfare that follow from this hypothesis.

Different Views of Animal Welfare

To understand how intensification affects animal welfare, we first need to be clear on what we mean by animal welfare. To do this, let us go back to the debate that arose when concerns were first expressed about the welfare of animals in intensive production.

* Professor in the Animal Welfare Program at the University of British Columbia, with a cross-appointment in the W. Maurice Young Centre for Applied Ethics

P.B. Thompson (ed.) *The Ethics of Intensification*,

The first major criticism of intensive systems came in the book *Animal Machines*, by the English animal advocate Ruth Harrison (1964). She described cages for laying hens and crates for veal calves, and she claimed that these systems are so unnatural that they cause animals to lead miserable and unhealthy lives. She went on to ask:

> How far have we the right to take our domination of the animal world? Have we the right to rob them of all pleasure in life simply to make more money more quickly out of their carcasses?

Subsequently, in *Animal Liberation*, Australian philosopher Peter Singer (1990) based his criticism of confinement production on the principle that actions should be judged right or wrong on the basis of the pain or pleasure that they cause. He claimed:

> There can be no moral justification for regarding the pain (or pleasure) that animals feel as less important than the same amount of pain (or pleasure) felt by humans.

In these and other quotations a key concern centred on words like "pleasure", "pain", "suffering", and "happiness". There is no simple English word to capture this class of concepts. They are sometimes called "feelings", but that term seems too insubstantial for states like pain and suffering. They are sometimes called "emotions", but emotions (by most definitions) do not include states like hunger and thirst. Perhaps the most accurate, if rather technical, term is "affective states", a term that refers to emotions and other feelings that are experienced as either pleasant or unpleasant rather than hedonically neutral.

In discussing intensive systems, however, some people put the emphasis elsewhere. A British committee that was formed to evaluate the welfare of farm animals concluded:

> In principle we disapprove of a degree of confinement of an animal which necessarily frustrates most of the major activities which make up its natural behaviour. (Brambell 1965)

Astrid Lindgren, the famous author of the Pippi Longstocking stories and a driving force behind animal welfare reform in Sweden, proposed:

> Let [farm animals] see the sun just once, get away from the murderous roar of the fans. Let them get to breathe fresh air for once, instead of manure gas. (Anonymous 1989)

And American philosopher Bernard Rollin (1993) insisted that we need:

> … a much increased concept of welfare. Not only will welfare mean control of pain and suffering, it will also entail nurturing and fulfilment of the animals' natures.

In these quotations, although affective states were often involved implicitly or explicitly, the central concern was for a degree of "naturalness" in the lives of animals: that animals should be able to perform their natural behaviour, that there should be natural elements in their environment, and that we should respect the "nature" of the animals themselves.

All of the above quotations reflected the views of social critics and philosophers, but when farmers and veterinarians engaged in the debate, they brought a different focus.

For example, one veterinarian defended intensive systems this way:

> My experience has been that … by-and-large the standard of welfare among animals kept in the so called 'intensive' systems is higher. On balance I feel that the animal is better cared for; it is certainly much freer from disease and attack by its mates; it receives much better attention from the attendants, is sure of shelter and bedding and a reasonable amount of good food and water. (Taylor 1972)

The British committee, cited above, reported hearing the common claim,

> … that the growth rate of an animal for meat or the egg production of a laying hen are the only reliable objective measures of their welfare. It is claimed that suffering of any kind is reflected by a corresponding fall in productivity. (Brambell 1965)

And the veterinary educator David Sainsbury (1986) proposed:

> Good health is the birthright of every animal that we rear, whether intensively or otherwise. If it becomes diseased we have failed in our duty to the animal and subjected it to a degree of suffering that cannot be readily estimated.

Here the primary emphasis is on the fairly traditional concerns of veterinarians and animal producers that animals should have freedom from disease and injury, plus food, water, shelter and other necessities of life – concerns that we might sum up as the basic health and functioning of the animals.

In these various quotations, then, we see a variety of concerns that can be grouped roughly under three broad headings, one centered on the affective states of animals, one on the ability of animals to lead reasonably natural lives, and one emphasizing basic health and functioning. The three broad areas of concern are not, of course, completely separate or mutually exclusive; in fact, they often go hand in hand. Harrison and Lindgren clearly believed that allowing animals to live a more natural life would make them more happy and healthy; Sainsbury clearly believed that unhealthy animals would suffer.

Nonetheless, the different areas of emphasis are sufficiently independent that the pursuit of any one will not necessarily improve animal welfare as judged by the other criteria. Housing animals outdoors may allow them to lead a seemingly natural life, but it can also expose them to predation, harsh weather, and diseases that might be better prevented in indoor facilities. Providing animals with ample amounts of palatable food may prevent the unpleasant affective state of hunger, but it can also lead to obesity and related health problems. Keeping animals in individual cages may prevent the spread of disease, but it may also prevent natural behaviour and result in frustration.

In trying to capture these ideas, both Appleby (1999) and Lund (2006) have used a graphic like the one shown in Fig. 12.1, which serves to summarize three points: that what people mean by animal welfare involves different concerns that can be grouped mainly under three headings; that these involve considerable but imperfect overlap; and that the pursuit of any one does not guarantee a high level of welfare as judged by the other criteria.[1]

[1] Here and previously (Fraser et al. 1997) I have based my analysis on an examination of samples of text. Ultimately we need actual studies of people's views of animal welfare to test whether these conclusions are sound, but some initial research tends to support them. For example, a study of Danish citizens' views on the welfare of pigs identified much the same elements, including concern over "physical harm", "avoidance of suffering", and that animals should be able to lead "a natural life" (Lassen et al. 2006). In a Dutch study, livestock producers tended to equate animal welfare with physical health, whereas consumers tended to emphasize "freedom to move and freedom to fulfill natural desires", without which an animal was believed to lead a "miserable life" (te Velde et al. 2002).

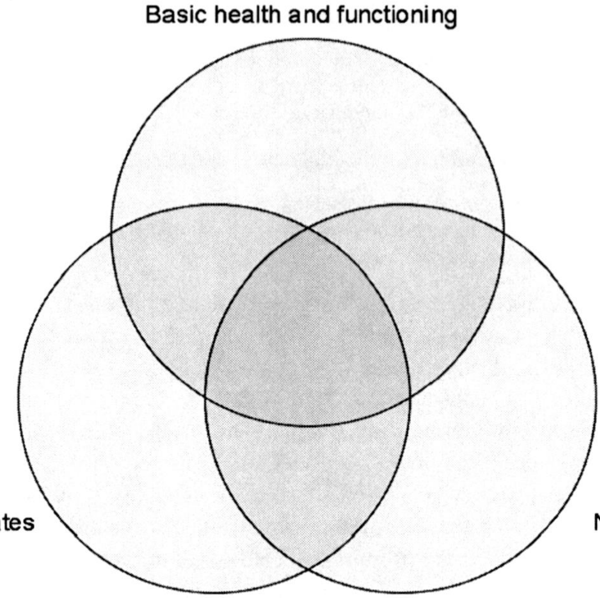

Fig. 12.1 Three conceptions of animal welfare (Adapted from Appleby 1999 and Lund 2006)

A Debate About Values

The different views of animal welfare do not necessarily involve disagreements about facts. An intensive animal producer might believe that welfare is good in a high-health confinement system because the animals are healthy and growing well; a critic might draw the opposite conclusion because the animals are crowded in barren pens and develop abnormal behaviour. The two parties may agree on factual issues such as the amount of space that is available and the incidence of disease. Their disagreement is about values – specifically about what they consider more important or less important for animals to have good lives.

But why should people hold such different views about what constitutes a good life for animals? Ultimately this is a question for social scientists, but I believe we can gain some insight by reviewing the debate that erupted during an earlier instance of intensification, namely the concentration of workers into cities and factories during the Industrial Revolution.

During the Industrial Revolution, the "factory system" became the predominant way of producing textiles and other goods throughout much of Europe, and the factories proved so efficient that traditional, hand production disappeared almost completely. Workers moved from villages and rural areas into cities; and instead of working at hand looms in their homes, people operated machinery in huge buildings. It was a profound social change, and it touched off an intense debate over whether the new factory system was good or bad for the people it affected (Bizup 2003).

On one side of the debate were critics who insisted that the factory system caused people to lead miserable and unwholesome lives. Critics claimed that the cities created cramped, unhealthy living conditions for the workers, and deprived people of contact with nature. The machines themselves caused many injuries, and (critics claimed) they often led to physical deformities because they placed an unnatural strain on the body. Perhaps worst of all, it was claimed that repetitive work with machines made the workers themselves like machines and led to an erosion of their human nature.

But the factory system also had staunch defenders. Instead of imposing strain on the workers, automation (the defenders claimed) relieved workers of much of the drudgery that manual handicrafts required. Far from being unnatural, the factory system represented a step in the natural progression from a time of human labour to a time when automation would make labour unnecessary. Moreover, the wise factory owner would take care to have healthy, happy workers because maximum productivity would not otherwise be achieved. In fact, the productivity of the system was seen as proof that the factories were actually well suited to human workers.

From this debate I believe we can build up a picture of the different values and world-views that lay behind the arguments. For simplicity, I have grouped these under two poles. The world-view of the anti-industrial critics might roughly be called Romantic/Agrarian,[2] and it reflected a set of values that we see extending from the Latin poetry of Virgil through to the Pastoralist and Romantic poets and painters of the 1600s to 1800s (Inge 1969). This world-view values a simple, natural life. It sees nature as an ideal state that we should strive to emulate. It values emotional experience and the freedom of the individual. And it looks back from industrial society to a Golden Age in the past when people lived in harmony with nature.

The world-view of the pro-industrialists was more a product of the Enlightenment when people looked to reason and science to replace what they saw as superstition and ignorance. This world-view involved two concepts that were relatively new to Western thought.

One of these was productivity. Adam Smith opened his book *The Wealth of Nations* (1776) by claiming that the quality of life in a nation depends on the goods that are available to supply the citizens with what they need and want. Increasing the productivity of the work force, and thus increasing the supply of goods, should therefore improve the lives of a nation's people. Thus the factory system, whereby automation and specialization lead to greater productivity, would ultimately lead to a better life.

The second idea was progress – the idea that human history moves irreversibly in the direction of improvement. As historian Sydney Pollard (1968) points out, belief in progress began with science, because in science each generation was seen as building on the work of earlier generations so that knowledge constantly improves. But during the 1700s the idea of progress took wing, and by 1800, in the words of Pollard,

[2] Although I have grouped Romantic and Agrarian for the purpose of this discussion, the two modes of thought have important differences and different histories as discussed in Fraser (2008b).

> firm convictions had been expressed about the inevitability of progress in wealth, in civilization, in social organization, in art and literature, even in human nature and biological make-up.

And a belief that change represents progress, and that we cannot "stand in the way of progress", has remained a common theme in Western thought ever since.

Thus, the Rational/Industrial world-view of the pro-factory voices was very different from the Romantic/Agrarian world-view of the critics. Instead of valuing a simple, natural life, it valued a life improved through science and technology. It viewed nature not as an ideal state that we should emulate, but as an imperfect state that we should control and improve. It valued rationality rather than irrational emotion, and the productivity of the well organized enterprise more than the freedom of the individual. And instead of looking back to a Golden Age of harmony with nature, it looked forward to a Golden Age in the future when progress through science and technology would lead to a better life.

The debate over human welfare during the Industrial Revolution has obvious parallels with the debate over animal welfare during the intensification of animal agriculture. In fact, much of the disagreement over animal welfare may well reflect the continued influence of the contrasting modes of thought. People who lean more toward a Romantic/Agrarian world-view will see a good life for animals as (primarily) a natural life, to be achieved by emulating nature through such means as free-range systems and access to the outdoors. They will emphasize the emotions of animals (are they suffering? are they happy?), and attach importance to their freedom. Such people are likely to see confinement systems as inherently incompatible with a high level of animal welfare, and they may look back to traditional, non-confinement systems as an ideal that we should try to return to. In contrast, those who lean more toward a Rational/Industrial world-view will tend to see a good life for animals as (primarily) a healthy life, to be achieved by preventing disease and avoiding the vicissitudes of nature. They will emphasize the rationality and scientific basis of the system (vaccines, scientifically formulated diets) more than the freedom of the individual animals, and they will see a high level of productivity as evidence that the animals are doing well. Thus, such people are likely to see modern, industrialized production systems as a form of progress that will ultimately lead to better animal and human welfare, and they may look upon older, agrarian systems as outmoded models that need to be improved upon.

Animal Welfare and Science

The early debates about the welfare of animals in intensive production systems were quickly followed by a body of scientific research designed to improve animal welfare. The question then arose: could such research resolve the value-based disagreements in the public debate by understanding animal welfare from the animal's "own point of view" (Dawkins 1990)? As a basis for the research, a seemingly logical first step was to agree on a "scientific definition" of animal welfare. Hence, in addition to their empirical research, scientists also engaged in a remarkable out-pouring of papers about the definition of the term. But as the various definitions were proposed, it became clear that the value-based ideas in the public debate were also intimately involved in the scientific debate about how to define animal welfare.

Some of the scientists tied the welfare concept firmly to the affective states of animals. Marian Dawkins (1988), for example, wrote that, "To be concerned about animal welfare is to be concerned with the subjective feelings of animals, particularly the unpleasant subjective feelings of suffering and pain". Ian Duncan (1993) agreed: "… neither health nor lack of stress nor fitness is necessary and/or sufficient to conclude that an animal has good welfare. Welfare is dependent on what animals feel."

Other scientists put the emphasis on various aspects of the animals' basic health and functioning. In some of his writing Donald Broom (1991) drew on the biological concept of "fitness" claiming that, "Poor welfare occurs in situations in which … there is reduced fitness or clear indications that fitness will be reduced." Stress physiologist Gary Moberg (1985) linked animal welfare to the theory of stress biology. He noted that the "stress" responses of the body are normally adaptive processes that help the animal to deal with challenges; however, if these responses are very prolonged and extreme, they can became "pre-pathological" in the sense of predisposing the animal to injury, disease or other problems. Moberg proposed, "that a prepathological state be used as an indicator of stress and risk to the animal's well-being." Other scientists tended to equate welfare with more traditional measures of agricultural productivity. For example, John McGlone (1993) proposed that, "an animal is in a poor state of welfare only when physiological systems are disturbed to the point that survival or reproduction are impaired."

Yet other scientists linked animal welfare to the ability of animals to live in a natural manner. For example, Marthe Kiley-Worthington (1989) proposed that for an animal not to suffer, "it is necessary over a period of time for the animal to perform all the behaviours in its repertoire because it is all functional" in the animal's life. In a similar vein, Susanne Waiblinger and coworkers (2004) noted that in many animal production systems, "major elements of natural behaviour cannot be performed, leading to welfare problems", and they called for more innovative production systems "in which all aspects of natural behaviour are taken into account." Biologists Chris Barnard and Jane Hurst (1996) linked their analysis of animal welfare to evolutionary theory by arguing that animals have been designed by natural selection not to preserve themselves by avoiding stress and hardship, but to "expend" themselves efficiently so as to achieve successful reproduction. They argued that any scientific conception of animal welfare must be based not on longevity, low levels of stress, or absence of suffering, but on allowing animals to live and expend themselves in the manner that is natural for the species.

The different conceptions of animal welfare gave rise to a wide range of scientific research that has helped to clarify issues and improve animal welfare. For example, studies on the affective states of animals have identified ways to prevent fear and pain; research on basic health has identified environmental factors that cause injury or predispose animals to disease; research on natural behaviour has identified housing and feeding systems that better correspond to the animals' natural manner of living. However, the science has not replaced the different value-based ideas about animal welfare with a single definition to which all scientists agree (Tannenbaum 1991; Sandøe and Simonsen 1992; Fraser et al. 1997). Instead, the different value-based positions – about what is most important or most desirable for animals – actually came to underlie the different definitions that the scientists proposed, and they stimulated different (often complementary) programs of research.

The Intensification of Animal Production

With that background on animal welfare, let us now be clear on what we mean by intensification.

During the past half century, world production of meat and other animal products has undergone a remarkable increase (Table 12.1). Production of poultry meat increased roughly eightfold, and production of pig meat roughly fourfold over 40 years. Production of meat from cattle, sheep and goats showed more modest increases, roughly doubling during the same 40 years, as did the world's human population.

Especially in the industrialized countries, these increases in production were accompanied by major changes in the structure and methods of animal agriculture. The various changes have tended to be lumped together under the heading "intensification", but for our purposes it is important to distinguish three elements.

One element was the consolidation of animal production into the hands of fewer and fewer owners. Table 12.2 shows the trends in Denmark and Canada, selected as

Table 12.1 World production, world exports, and percentage exported, for four categories of meat from 1961 to 2001 (FAO 2004a)

Source of meat	1961	1971	1981	1991	2001
World production (1,000 t/year)					
Poultry meat	8,911	15,657	27,386	42,939	71,414
Pig meat	24,702	39,345	52,903	71,784	92,071
Bovine meat	28,737	39,386	47,581	56,278	59,149
Mutton/goat meat	6,026	6,934	7,585	9,811	11,449
World exports (1,000 t/year)					
Poultry meat	303	594	1,900	2,923	9,359
Pig meat	1,092	2,025	2,788	4,618	7,752
Bovine meat	1,658	2,886	4,692	6,940	7,431
Mutton/goat meat	487	715	940	848	874
Percentage exported					
Poultry meat	3.4	3.8	6.9	6.8	13.1
Pig meat	4.4	5.1	5.3	6.4	8.4
Bovine meat	5.8	7.3	9.9	12.3	12.6
Mutton/goat meat	8.1	10.3	12.4	8.6	7.6

Table 12.2 The number of farms raising major meat-producing species in Denmark and Canada

	1970/71	1980/81	1990/91	2000/01	% Change per year[d]
Denmark[a]					
Chickens	68,900	27,569	13,764	5,676	–3.1
Pigs	120,370	67,708	29,903	13,231	–3.0
Cattle	103,465	61,310	36,432	23,031	–2.6
Sheep	4,835	4,182	6,266	3,241	–1.1
Canada[b]					
Chickens[c]		55,017	24,305	10,875	–4.0
Pigs		55,765	29,592	15,472	–3.6
Cattle/calves		185,073	145,747	122,066	–1.7
Sheep		12,905	13,114	13,232	0.1

[a]Danmarks Statistik (2004). Chapter 10: Livestock, Table 10.2: Livestock holdings at structural surveys.
[b]Statistics Canada (2005).
[c]Broiler chickens and Cornish hens.
[d]% change per year is calculated as the decline per year (the number of farms in the first year, minus the number in the final year, divided by the number of years) expressed as a percentage of the number of farms in the first year of the data set.

two countries with large animal industries. Chicken and pig production, which showed a large increase in total production during this period, also showed a pronounced decline in the number of farms raising these species. For cattle the decline was slower, and there was no consistent change in the number of farms raising sheep.

A second element was the confinement of large numbers of animals into buildings or feedlots. This occurred mainly for the species or production phases that involve feeding grain and other concentrated diets. Thus, pig, poultry and egg production (which are largely grain-based) and veal-calf production (which uses concentrated milk- or grain-based diets) moved largely into indoor facilities. The change was much less pronounced for predominantly forage-fed animals. For example, many beef cattle in North America, although concentrated in large outdoor feedlots where they are finished on grain-based diets for their last few months, are raised for much of their lives in traditional grazing systems, and most sheep and goats continue to be raised in traditional, extensive systems (Fraser et al. 2001).

A third element involved a range of changes that might be grouped together under the heading of "cost-cutting". In the indoor units, the amount of space provided per animal was often very limited. Many pregnant sows and veal calves, for example, were kept in stalls too narrow to allow them to walk or turn around, and many laying hens were kept in cages that were so small as to allow only very limited movement. Labour costs were also curtailed: as herd and flock size increased, the amount of staff time did not increase proportionally, and in many cases the amount of labour per animal was greatly reduced. Amenities such as bedding were often minimized or eliminated.

For the purposes of this paper, I will use "intensification" to refer to these three changes: the consolidation of animal production in fewer and fewer hands, the concentration of animals into buildings and feedlots, and the cost-cutting that resulted in limited freedom of movement, limited staff time per animal, and the minimizing or elimination of certain amenities. In making these distinctions we also need to be clear about the term "confinement". In some cases, "confinement" has been used simply when animals are confined inside buildings or feedlots, but in other cases the term is used for the close restriction of movement that occurs when animals are kept in very small holding units such as cages for hens and stalls for sows. An indoor aviary for laying hens is a confinement system in the first sense because the birds are kept indoors, but not in the second sense because the birds are free to move around the building. For clarity, I will use "confinement systems" to refer to production systems that involve confinement in both senses, and indicate the specific sense if one or other of the more narrow meanings is intended.

Intensification and Ethical Conflict

Perhaps it was a coincidence of timing, but during the same half-century when animal agriculture was undergoing massive intensification in the industrialized countries, a major shift was underway in attitudes towards animals in Western society. In a trend dating back at least to 1700, animals have been receiving increasing amounts of attention and sympathy in literature, the visual arts and philosophy (Harwood 1928). The change in attitudes may have resulted in part from growing scientific knowledge about animals, which has tended to narrow the gap that people perceive between themselves and other species (Fraser 2001b). Major developments in the twentieth century may also have fueled shifting attitudes. In particular, the nature of human exposure to animals changed as people moved from rural to urban living, and the media made the lives of wild animals accessible to people as never before. Whatever the root causes, the latter half of the twentieth century saw a pronounced increase in attention to animal issues and concern over animal welfare. As a result, all institutionalized forms of animal use – in science, entertainment, wildlife management and elsewhere – were subjected to critical scrutiny.

To some extent, however, the use of animals in agriculture was shielded from such scrutiny in the West by two powerful moral ideas. One is a strong positive attitude toward the diligent care of animals, inherited in part from the Bible. In the pastoralist culture that gave rise to the Bible, the raising of domestic flocks and herds was a major economic activity. Hence, it is not surprising that this culture legitimized the ownership and use of animals. However, pastoralism also required that humans provide appropriate care for animals. Hence, it is also unsurprising that the culture attached great value to animal care. For example David, who was chosen by God to become a great king of his people, began his career as a shepherd who showed exceptional courage in protecting his family's sheep. The sign that Rebecca had been chosen as the wife of Isaac and ancestor of her nation was her offer to water the camels of a stranger. Indeed, a shepherd protecting a flock of sheep was such a positive image that it was used as a common metaphor for divine goodness. Thus, the biblical culture gave rise to an ethic that legitimized certain uses of animals and simultaneously attached great value to diligent animal care. Against this background, the production of animals has been seen as an acceptable – even virtuous – activity, as long as appropriate care was involved (Preece and Fraser 2000).

A second important moral idea is the veneration of farmers and farm families living in a harmonious "agrarian" relationship with the land. As literary scholar Thomas Inge (1969) notes, there is a long-standing belief in Western thought that agrarian living brings out the best in humankind. As early as the fourth century BC, Aristotle had proposed that, "The best common people are the agricultural population, so that it is possible to introduce democracy as well as other forms of constitution where the multitude lives by agriculture or by pasturing cattle". In Rome, writers such as Cicero, Cato the Elder and Horace extolled agriculture as the noblest of occupations and the most likely to foster virtuous conduct. Within English literature, Inge argues, "the virtuous, simple life in the country became ... one of the most common thematic motifs". In America Thomas Jefferson took up the theme in 1781 in his *Notes on the State of Virginia*, claiming that, "Those who labour in the earth are the chosen people of God, if ever he had a chosen people..." (Inge 1969).

Animals played a less central role in agrarianism than in pastoralism, but they were still a crucial component. As philosopher Paul Thompson (1998) notes, farm animals were an integral part of the ecology and economy of the agrarian farm. They also played a key role in moral education, because children often learned responsibility by caring for animals. And animals on traditional farms, like the agrarian family itself, were seen as living natural and wholesome lives. Thus, the production of animals was seen as an acceptable – or even virtuous – activity, as long as it happened within an agrarian context.

The intensification of animal production fell afoul of both of these cherished moral ideas. Because intensification involves far fewer, larger, and more specialized facilities, it was perceived as contributing to a decline of family farming, and the use of seemingly industrial buildings and equipment clashed with the traditional image of agrarian life. Intensification also appeared, at least to outsiders, to be at odds with the ideals of attentive animal care. Instead of a good shepherd diligently seeking a lost lamb, skeptics pictured livestock producers jamming huge numbers of animals into inadequate cages and stalls in order (quoting Ruth Harrison, above), "to make more money more quickly out of their carcasses."

In summary, the intensification of animal production occurred at a time of growing public attention to animals and growing concern for their welfare. It also clashed with two influential images of animal agriculture – images that had arguably given the raising and killing of animals much of its moral legitimacy in the West. Hence, in the case of animal production, the intensification of agriculture was seen not merely as a controversial and perhaps imprudent or unsustainable shift in food production; it was seen as an affront to cherished ideals. This almost certainly helps to explain why in some quarters it has triggered a reaction of moral outrage (Fraser 2001a).

The "Standard Critique"

The criticisms and condemnations of intensive animal production have followed a pattern that has been repeated so faithfully – in books, broadcasts, Websites and other media – that we might call it the "Standard Critique" of intensive animal production. It includes several related claims.

One claim is that the intensification of animal production is a result of large corporations taking over family farms. For example, the website of Animal Place (undated) states, "Over the past 50 years, animal agriculture has evolved from small, family farms to large corporate factory farming systems" and that these "are built upon the cutthroat attitude of increasing profit margins at all costs ...". The popular animated film *The Meatrix* claims, "In the mid twentieth century, greedy agriculture corporations began modifying sustainable family farming to maximize their profits at great cost to both humans and animals" (Sustainable Table and Free Range Studios 2003). Critics often attributed confinement production systems to the influence of corporate owners. For example, Edward Dolan (1986), in *Animal Rights*, claimed that in "the world's most advanced countries", the family farm is "being replaced by large companies that ... are replacing the old farm ways with modern production methods." In *Vegan: The New Ethics of Eating*, Erik Marcus (1998) wrote that in the 1980s, "big corporations stepped in and took over the pig industry with the same large-scale systems applied to poultry", and, "with the decline of the family farm, animals that used to be cared for with kindness and a general regard for their welfare now live and die in unconscionable conditions."

A second claim is that traditional animal care values have been displaced by greed for profit. In *Old MacDonald's Factory Farm*, C.D. Coats (1989) stated, "Now humane treatment is seen as unnecessary, irrelevant, and in conflict with the maximization of profit". In describing "intensive confinement factory" production, Anna Sequoia (1990) claimed that, "Absolutely no consideration is given to the comfort of these animals, except to keep them alive in large enough numbers to make the animal factories profitable." Edward Dolan (1986) claimed that "the natures, welfare, and comfort of the animals are totally ignored for the sake of production methods that seek the greatest profit possible at the least possible cost in housing and care." John Robbins (1987), in *Diet for a New America*, tied this change of values to the influence of corporations, claiming that "the behemoths of modern agribusiness seek profit without reference to any ethical sensitivity to the animals in their keeping".

A third claim is that the methods of intensive production have horrible effects on animal welfare. For example, in *The Price of Meat,* author Danny Penman (1996) states, "Whether they are battery chickens in their cages or pigs in sow stalls, all experience the same mental anguish that would drive many humans to suicide." The Humane Farming Association (undated) speaks of animals in "factory" farms leading lives "characterized by acute deprivation, stress, and disease." Sequoia (1990) claimed that, "Five billion animals suffer from birth to death on factory farms each year."

Thus, the Standard Critique portrays the intensification of animal production as a process whereby corporations have replaced family farms and substituted industrial methods for traditional farming methods; it claims that profit motivation has replaced the traditional animal care values that were present in family farming; and it claims that all of this has had terrible consequences for animal welfare.

Doubts About the Standard Critique

The Standard Critique appears to have exerted great influence on popular beliefs about animal production, but is it an approximately correct analysis of how intensification came about? Unfortunately, most of the research needed to understand the intensification of animal production has yet to be done, but even at this stage we can see that the Standard Critique does not fit with some key facts.

One problematic claim is that the intensification of animal production is closely linked to corporations replacing the family-owned farm. Research is needed to assess this claim, but it appears that although intensification has occurred throughout the industrialized countries, corporate ownership has become the norm only for certain commodities in certain countries. In the United States of America, the majority of egg and poultry production is now controlled by a handful of corporations. In Canada, however, although egg and poultry production has also shifted to fewer, larger, indoor units, the individual producer remains the dominant player. This is likely because of a supply-management system that has kept profit per bird much higher than it is in the United States (Hunton and Fairhurst 2002). Similarly, the past two decades have seen the appearance of huge, corporately owned swine production companies in the United States. However, these are seen as an aberration in many other countries where swine production has nonetheless become highly intensified. In fact, the wholesale replacement of family-owned farms by large, corporately controlled units appears to have occurred mainly in two areas of the world – in some sectors in the United States of America and in some countries of the former Soviet Union. In both these cases, it might be argued that strong adherence to a particular economic system – market capitalism in the United States and collectivism in the former Soviet Union – was sufficient to displace the traditions of rural life. Elsewhere it appears that the intensification of animal production occurred mainly in a climate of individual and family ownership, and that much of the increase in farm size is a product of owner-operated units becoming progressively larger. Thus, the view that intensification is closely linked to corporate ownership is true of only certain sectors in certain countries.

Correlating confinement housing with corporate ownership is almost certainly incorrect. For one thing, the timing is wrong: most of the confinement methods in use today were becoming standard technology during the 1960s and 1970s, well before large, corporately owned units became common. Moreover, confinement technology predominates in many industrialized countries where individually and family-owned units remain the norm, with the exception of special cases such as Norway where subsidization allows small units and more traditional methods to remain viable. In fact, confinement methods are often staunchly defended by producers operating individually and family-owned enterprises (e.g., Kuehn and Kahl 2005).

Is it true that today's animal producers have undergone a major shift away from traditional animal care values? This question requires social science research into the values of animal producers. In the absence of such research, critics have approached the issue more with rhetoric than with investigation. Critical works such as *Diet for a New America* (Robbins 1987) and *Animal Liberation* (Singer 1990) provide quotations from modern animal producers illustrating extreme callousness toward their animals, and because these books give no quotations from the other end of the moral spectrum, they leave the impression that such callousness is typical. On the other hand Kolkman (1987) provides ample quotations from modern animal producers who espouse traditional values of stewardship and care for animals.

Thus, a wide range of values is clearly present, but has there been a change on net away from traditional animal care values? This is certainly possible. As animal production moved towards fewer, larger units, the producers who remained in business may well have, on average, different attitudes from those who left; and owning a large herd of dairy cows may engender somewhat different attitudes towards animals than owning a small herd (Waiblinger and Menke 1999). Nonetheless, the small amount of relevant research has found that people engaged in commercial animal production have a wide range of attitudes toward animals – some of them very positive – and that positive attitudes are correlated with the productivity of the animals (Hemsworth and Coleman 1998). Pending more thorough study it seems plausible to conclude that producers' attitudes toward animals cover a wide range, that they vary from the callous to the caring, and that some or many producers today continue to espouse strong animal care values although experiencing serious constraints on their ability to act on those values in today's world (Fraser 2006a).

An Alternative Hypothesis[3]

If the intensification of animal production is not principally the result of corporate takeovers, excessive profit taking and the disappearance of animal care values, then what did cause it? The amount of investigation and analysis needed to answer this question could fuel a small industry of research in history, economics and rural sociology, but in the absence of such research, here is an alternative hypothesis to account for the intensification of animal production.

---

[3] In this chapter I am dealing with the intensification of production that occurred at the level of the farm. With this intensification came changes in many other segments of the animal-based industries. Examples include long-distance transport of animals, slaughter in centralized plants killing thousands of animals per day, and genetic selection (mostly by breeding companies) for production traits such as rapid growth, high milk yield, and heavy body weight. All of these have had important effects on animal welfare, but are not the subject of this chapter.

First, we must acknowledge that a variety of contributing factors were no doubt involved. Demographics likely played a role: in particular, the expansion of industrial economies meant that agriculture had trouble retaining the work force needed to operate the older methods of animal production, and confining animals indoors allowed the use of automation and worker convenience to reduce labour requirements. Policy decisions were involved, as many governments promoted a shift to larger, specialized units as a way to produce a secure supply of inexpensive food and to improve the lot of low-income farmers (Thompson 2001). New animal production technology doubtless played a role, especially new vaccines and other medications that allowed large numbers of animals to be raised successfully in small spaces. And cultural values no doubt contributed: to some farmers in the 1950s and 1960s it must have seemed modern and progressive to use machinery and other hardware to eliminate manual tasks.

However, I would hypothesize that two common forms of twentieth century technology, namely refrigeration and road transportation, played a role that has been largely overlooked. During much of the nineteenth century, the main methods of transporting animals for long distances were the rail and water systems. Because these were accessible to only a certain fraction of farms, many animals could not easily be sent to concentrated slaughter facilities far from their point of origin. Thus, many animals would have been slaughtered on the farm or in local facilities. Although certain products such as smoked ham and salted pork were sufficiently preserved to be transported for sale elsewhere, most animal products, being highly perishable, must have been sold fairly close to the point of production through small, local butchers and dairies. Hence, farmers producing animal products would often have found themselves in competition with a modest number of other local producers working under the same conditions of weather, feed availability and labour costs.

By the late nineteenth century, however, rail transport allowed the slaughter industry to become more concentrated. This trend was greatly increased in the twentieth century by the vast expansion of roads and vehicles which allowed live animals to be shipped from virtually any farm to distant slaughter plants. In addition, the development of refrigeration, deep freezing and related technologies allowed meat and other animal products to be shipped to markets in different regions, different countries, or different continents.

These developments would have allowed the slaughter and processing industries to become concentrated in fewer and fewer companies, because a single plant could source animals and sell products over a very large geographic area. With a vast number of producers selling to a small number of large processors, we might expect that market competition would lead to very low levels of profit for the farmer until some further development (such as cooperative marketing or consolidation of production in fewer hands) reduced the pressure of competition.

I will argue (below) that pressures created by periods of low profit played a key role in the intensification of animal production and had important effects on animal welfare. First, however, let us test whether the two key assumptions – increased movement of animal products and periods of low profits – fit with the facts.

Was intensification accompanied by increased movement of animal products? The idea is difficult to test directly because most of the transportation was probably within national boundaries and not subject to data collection. However, if we consider export as an indicative tip of a much larger iceberg, then available data do indicate that the movement of animal products was increasing rapidly during the half century when the intensification of animal production was in progress. As shown in Table 12.1, the rate of growth in exports of meat during the period 1961–2001 far exceeded the rate of growth in the production of intensively produced animals. For poultry meat the percentage exported rose rapidly from 3.4 percent in 1961 to 13.1 percent of a much larger output by 2001. For pig meat and bovine meat, the percentage exported roughly doubled during this time. In contrast, for mutton and goat meat, products of animals that were less subject to intensification, there was little change in the percentage exported.

Was intensification accompanied by low profits? Data need to be examined from many commodities and countries, but available figures from the United States of America provide striking examples. According to data from Dr. John D. Lawrence of Iowa State University (Table 12.3), profit from farrow-to-finish pig production in the United States averaged about US$21 per animal (roughly US$0.20–0.25 per kilogram) in 1974–79 and then declined to about US$7 per animal in the 1980s and US$4 in the 1990s, with years of loss mixed with years of modest profit. If we take inflation into account, the fall in profit would be even more drastic than these figures imply. The United States chicken meat industry, which experienced consolidation earlier than the swine industry, was also earlier to experience near-zero profit levels. In 1970–79, United States chicken production generated profits in only five of the ten years, with an average profit of only about US$0.02 cents per kilogram during the decade. The industry underwent massive consolidation to the point that about half of production was controlled by ten companies by 1980 and by five companies by the mid-1990s (Thornton 2003). As this occurred, profits increased substantially, and production has been profitable every year since the mid 1980s. Egg production in the United States underwent a similar process (Table 12.3).

Periods of low and fluctuating profit could account in several ways for the intensification of animal production.

First, low profits must have been a powerful factor in the move toward larger farms. With a substantial profit per animal, a family could make a living from a relatively small unit, but with low profit per animal, such units would no longer generate enough income to support a family; hence, producers would be forced to expand or find other employment. For example, from data in Table 12.3 we can calculate that in the 1970s a family-run unit with 120 sows and an annual production of 2,000 pigs would, on average, generate an annual profit of about US$42,000 – a good living for a family at the time. In the 1990s, the profit from such a unit would have been only about US$8,000 – more a hobby than an income, and in bad years a hobby that few could afford. Thus, farmers who wanted to continue producing animals for a living had little choice but to expand their operations considerably.

Second, low and fluctuating profits may also have encouraged a move toward indoor systems because of a need to reduce losses and production costs. Indoor

Table 12.3 Historical data on profit from three types of animal production in the United States of America

Year	Eggs: Net returns in US$0.01/dozen[a]	Chicken: Net returns in US$0.01/kg[a]	Pigs: Profit from farrow-to-finish operations in US$1/head[b]
1967		−1.1	
1968		4.0	
1969		6.4	
Decade		3.1	
1970		−1.1	
1971		−0.7	
1972	−2.8	−0.2	
1973	6.2	5.7	
1974	−0.4	−4.4	1.46
1975	1.0	12.8	34.53
1976	10.0	2.0	23.89
1977	3.8	0.9	16.70
1978	1.7	9.7	36.33
1979	3.1	−2.4	13.85
Decade	2.8	2.2	21.13
1980	−3.5	−3.3	−4.13
1981	0.4	−11.9	−5.03
1982	3.8	−6.2	28.96
1983	3.3	−3.7	−0.65
1984	8.9	10.3	−2.24
1985	1.2	13.4	−0.20
1986	7.0	27.9	23.36
1987	0.2	4.4	34.29
1988	−5.0	12.3	0.55
1989	15.2	15.8	−3.88
Decade	3.2	5.9	7.10
1990	16.7	12.1	29.92
1991	12.8	6.6	21.99
1992	1.7	7.3	3.83
1993	8.6	14.1	12.71
1994	3.5	12.8	−8.28
1995	8.8	16.6	0.87
1996	13.1	11.9	9.99
1997	11.6	12.8	13.61
1998	13.1	31.5	−27.98
1999	5.9	26.0	−17.29
Decade	9.6	15.1	3.94
2000	9.0	20.2	11.89
2001	6.7	25.7	13.67
2002	4.4	17.2	−16.21
2003			−6.34
2004			22.55
Decade	6.7	21.0	5.11

[a]USDA (2004) (Table 050, Eggs: net returns, and Table 091, Young chicken: net returns (converted from cents per pound).

[b]Data from John D. Lawrence, Iowa State University.

production, despite the high capital cost, was a way to reduce operating costs by using automation and worker convenience to reduce labour, and it probably lowered feed costs by keeping animals warm in cold weather. In many cases, indoor production also helped to reduce some of the traditional losses resulting from disease: the cage for laying hens was in part a way of separating birds from pathogens in soil and excreta, and enclosed barns allowed producers to create biosecurity barriers to prevent the introduction of diseases. Indoor systems also prevented deaths (especially of young animals) caused by predators and harsh weather. Moreover, for farmers who did not acquire proportionally more land as they increased their herd or flock size, confining the animals in buildings or feedlots may have been more or less necessary in order to raise many more animals on the same land base.

In some cases, low and fluctuating profits may also have encouraged the integration of farm production into some form of corporate structure. Linking numerous farms in a company that also produced feed and processed meat presumably helped to achieve economies of scale. Perhaps more importantly, with near-zero profit at the farm gate, profit might still be made at other points in the production process; hence, producing chickens within a corporate structure likely remained profitable even when independent chicken farms were not. However, whereas the expansion of herds and flocks was virtually a necessity to generate a family income, corporatization was an option that was followed only in some cases.

In addition to the above effects, low profits may have had their greatest effect on the third element of intensification: the cost-cutting that accompanied the move toward larger units and indoor production. One aspect of cost-cutting was the limited amount of space allowed per animal. In egg production in the United States, for example, a space allowance of only about 300 cm^2 per bird was relatively common. This space allowance maximized profit under certain market conditions, but it was so small that it reduced average survival and egg production per bird (Adams and Craig 1985; see also Bell et al. 2004). Similarly, staff time per animal was drastically reduced. This in turn led to changes in the physical environment. For example, the use of slotted floors and the elimination of straw from much pig production was largely an attempt to the avoid the labour needed for manual removal of manure, and the gestation stall was a means of eliminating the daily labour needed to move sows from group pens into feeding stalls for individual feeding.

In summary, the alternative hypothesis proposes that developments in the twentieth century, notably in transportation and food preservation methods, allowed much greater movement of animals and animal products, and led to consolidation of the food processing industry; that the resulting increase in competition among animal producers caused periods when the producers received very low profit per animal; and that these periods of low profit were a major factor contributing to the shift towards larger units and indoor production, and also necessitated cost-cutting in factors such as space, staff time and other amenities.

Intensification and Animal Welfare

How has the intensification of animal production affected the welfare of the animals? If we think of intensification as consisting of three broad elements – consolidation of ownership, confinement of animals to buildings and feedlots, and the cost-cutting that reduced space and other amenities – then we can ask how animal welfare has been influenced by each of these in turn. Indeed, if we are to make effective recommendations to improve animal welfare, it will be important to understand which aspects of intensification had which effects.

First, has the shift from small farms to larger farms, or even to corporately owned farms, been bad for animal welfare? Lacking empirical evidence, we can only propose tentative answers. It seems plausible that the quality of animal care might decline in very large units, for example if the staff are wage-earners who have no stake in the enterprise, or if important decisions are made by executives who are not in contact with the animals. On the other hand, very small units may lack the capital, specific knowledge and access to specialized services that a larger unit can provide. Thus, if we could plot "average" animal welfare (however this might be conceived) against farm size, the line might be slightly hill-shaped, climbing at first as the specialist takes over from the generalist-level skill present on the small mixed farm, and then declining with very large units if key decisions are made by people who do not care about the animals or are not in contact with them. However, given the multitude of factors that influence animal welfare, we might suspect that the hill, if it exists, would not be very steep.

Second, does the confinement of animals in buildings or feedlots by itself lead to reduced animal welfare? The answer will depend greatly on the different views of animal welfare discussed above. For those whose thinking is influenced more by a Romantic/Agrarian world-view, confinement of animals in human-made, indoor environments is inherently bad for their welfare because the environment is unnatural and the animals' ability to carry out their natural behaviour is likely to be limited. In contrast, people who are more influenced by a Rational/Industrial world-view may tend to down-play these problems and point instead to how indoor environments can lead to better health through protection from harsh weather and predators, the ability to eliminate certain pathogens, and the ability to treat animals easily for diseases and parasites. Thus, on the specific question of keeping animals indoors versus partly or wholly outdoors, we are likely to see disagreement between those who emphasize different elements of animal welfare.

However, when we consider the third element of intensification – the cost-cutting that often went along with larger units and indoor housing – then we find much more agreement about effects on animal welfare. For example, whether they emphasize affective states, natural behaviour or basic health and functioning, all informed participants acknowledge that animal welfare can be compromised if animals have inadequate space. Specifically, with too little space animals are likely to be uncomfortable and frustrated (affective states); they will not be able to move or even rest and sleep in a normal manner (natural behaviour); and in extreme cases they will show reduced survival and growth (basic health and functioning).

Much the same applies to cost-cutting in other areas. Skillful animal handling, and the prompt detection and treatment of disease, are universally recognized as important for animal welfare, but these are likely to be sacrificed if there is not sufficient profit to hire, train and retain good workers. Some of the labour-saving changes to the animals' physical environment have well recognized effects on animal welfare. In pig production, for example, slotted floors and under-floor manure pits, which were used to mechanize the handling of manure, contribute to high levels of air-borne ammonia which affect the comfort, health and behaviour of the animals. Similar arguments can be made for cost-cutting in other areas such as the amount of bedding used for dairy cows, the maintenance of equipment, the quality of surfaces for resting and walking, and so on.

In short, although reasonable people may disagree about how large farms and indoor housing affect animal welfare, there is little disagreement that the cost-cutting features of intensification – cutting back on space, staff time, the quality of the animals' environment – constitute threats to animal welfare by any definition.

Approaches to Promoting Animal Welfare

In proposing how to promote the welfare of farm animals, proponents of the Standard Critique have generally offered two main options: eliminating the consumption of meat as a way of avoiding the products of intensive animal production, or returning to the type of agriculture that preceded intensification. Over the past 30 years these have provided satisfying personal-choice options for certain consumers and for those farmers who embrace an agrarian manner of living.

However, data on the current production and consumption of animal products show that something more is needed. In the industrialized countries, per capita meat consumption appears to have plateaued at a very high level, and consumption in the less industrialized countries continues to increase (Table 12.4). As a result, we are now entering a century with unprecedented yet still increasing levels of meat production, the majority of which is occurring in the less industrialized nations. Although there has been growth in alternative production systems such as organic and free-range (Vaarst et al. 2004), much of the increase in global animal production is almost certainly due to increases in medium- to large-scale confinement production. Thus, there is a pressing need not only for personal-choice options, but for social-policy solutions that will promote animal welfare in a world where vast amounts of animal products will continue to be eaten, and where much animal production will continue to follow intensive production models. The alternative hypothesis suggests several courses of action.

First, if it is true that ethical values regarding animal care have not disappeared, but rather that economic constraints limit the ability of farmers to act on those values, then one action could be a project for the social sciences: to identify producers who

Table 12.4 Supply of four types of meat in developing countries, developed countries, and the world, 1961–2001, expressed in kilograms per person per year source (Including division of developed and developing countries): FAO, 2004b

Source of meat	1961	1971	1981	1991	2001
Poultry meat					
Developing Countries	1.0	1.5	2.7	4.2	7.8
Developed Countries	6.7	10.5	15.3	20.1	24.3
World	2.9	4.1	6.0	8.0	11.3
Pig meat					
Developing Countries	2.1	4.0	5.6	8.4	11.4
Developed Countries	20.5	25.3	28.8	29.1	28.0
World	8.0	10.2	11.6	13.3	15.0
Bovine meat					
Developing Countries	4.3	4.0	4.7	5.1	6.1
Developed Countries	19.9	26.1	26.6	27.0	21.4
World	9.3	10.4	10.4	10.3	9.4
Mutton and goat					
Developing Countries	1.2	1.2	1.3	1.5	1.7
Developed Countries	3.5	3.4	2.7	2.8	2.0
World	1.9	1.8	1.6	1.8	1.8
Total of the four					
Developing Countries	8.6	10.7	14.3	19.2	27.0
Developed Countries	50.6	65.2	74.4	79.0	75.7
World	22.1	26.5	29.6	33.4	37.5

maintain strong animal care values, and to understand from them what changes would free them to raise animals in ways they consider appropriate. Such an understanding might provide a basis for building consensus between producers and consumers about how improved animal welfare could be achieved.

Second, if inadequate profits do seriously limit the ability of producers to use practices that promote animal welfare, then a large part of the solution will need to be economic. Specifically, producers will need to be protected from market pressures that force them to cut back on space, bedding, staff time, and other factors that are important for animal welfare. Examples of such remedies include: (1) product-differentiation programmes that provide premium prices for products produced according to specific standards, (2) government programmes to help producers adjust to animal welfare standards, perhaps modeled after incentives to encourage conversion to organic methods, and (3) purchasing agreements whereby corporate customers (chain restaurants, retail chains) agree to pay higher prices in return for guarantees of animal welfare standards (Fraser 2006b). On the other hand, if and as profitability returns to an industry (as appears to have happened in chicken and egg production in the United States), then it is reasonable for society to expect the industry as a whole to meet appropriate welfare standards.

Moreover, if it is true that the forces of marketing and trade have been the primary drivers of intensification, one strategy may be to mold these forces in ways that promote animal welfare. For example, the European Union provides trading opportunities to its member countries but also requires that they conform to agreed animal welfare standards (Stevenson 2004). The European Union has also indicated a willingness to create opportunities for developing countries to supply Europe with products from animals produced according to certain animal welfare standards (CEC 2006). International animal welfare guidelines, such as those of the World Organization for Animal Health, could be incorporated in agreements between cooperating nations as a basis for trade in animal products (Thiermann and Babcock 2005; Fraser 2008a). Certainly, as long-distance trade in animal products increases further through trade liberalization, we need to ensure that this trend does not lead to a new phase of near-zero profit which further constrains the ability of producers to act in ways that favour animal welfare.

Third, if the problems of animal welfare have less to do with keeping animals in buildings and more to do with the cost-cutting in space, staff time etc. that evolved with it, then we need a different focus in advocating changes to production methods. The pressure for animal welfare reform should not focus mainly on eliminating indoor systems but on identifying animal welfare problems in those (and other) systems and proposing remedial actions. This is a much more complex programme than simply calling for an end to large farms and indoor production, and it would also extend to other elements of animal production such as animal breeding, animal transport, and training of workers and veterinarians. Such an approach could create opportunities for animal advocates and animal producers to pursue common goals. Some progress has been made in this regard. In the Canadian province of Alberta, for example, a cooperative programme involving the animal protection movement and livestock producers has resulted in training, inspection, enforcement and research, which both groups support (AFAC undated).

Finally, we may need to change our vision of what constitutes a good animal producer. The ideal envisaged by the Standard Critique draws on the occupational models of agrarianism and pastoralism, and seeks to replace intensive animal producers with smaller-scale farmers using more traditional production methods. Undoubtedly there are producers who would embrace such a vision, but many would not. However, during the last century the flourishing of the professions – medicine, teaching and others – provided a different occupational model. To be part of a profession involves, among other things, acquiring requisite knowledge and skills through training and supervision, recognizing (and sometimes helping to shape) accepted standards of professional behaviour, conforming to those standards and helping to ensure that all members of the profession do the same. Perhaps a model of professionalism in animal production would provide an alternative vision that many agricultural producers might find appealing and achievable, and would support a move toward greater attention to animal welfare in intensive production.

Acknowledgments This chapter is a revised version of a paper I wrote for the *FAO Readings in Ethics Series* (Fraser 2005). Portions of the text are also based on other publications listed below as Fraser (2006a) and Fraser (2008b). I am grateful to the publishers for permission to use sections of those texts. I am also grateful to the Food and Agriculture Organization of the United Nations (FAO), the Natural Sciences and Engineering Research Council of Canada, and the Social Sciences and Humanities Research Council of Canada for supporting the research; to Camilla Ryberg for diligent research assistance; to John D. Lawrence for kindly providing data on profits in swine production; and to Rod Preece and to many colleagues at the FAO and the University of British Columbia Animal Welfare Program for helpful discussion.

Chapter 13
Re-thinking the Ethics of Intensification for Animal Agriculture: Comments on David Fraser, Animal Welfare and the Intensification of Animal Production[1]

Peter Sandøe*

In a recent essay published in the prestigious series of readings in ethics issued by the Food and Agriculture Organization of the United Nations David Fraser discusses the relationship between animal welfare and modern, intensive animal production. His main view is that it is possible to uphold a decent level of animal welfare within intensive animal production and that the best way to promote animal welfare is to reform rather than reject intensive animal production.

Fraser is aware that other people hold credible views on this subject; he is also aware that his own views differ from those of influential authors who, approaching the matter from the perspective of animal welfare or animal rights, often criticize intensive animal production. Indeed, it should be noted that Fraser's chapter in the present volume seems to have anticipated many of the criticisms that I present here and therefore, as far as I can see, has moderated a number of the claims made in the FAO booklet. My comment here is offered to highlight some of the important points on which these two statements differ. The original FAO Ethics Paper is, in fact, largely an argued rejection of the perspective and conclusions of critics of intensification of animal production.

Specifically, the FAO document criticizes the following three widely believed propositions: (a) there is a strong link between intensive animal production and problems with animal welfare; (b) the development and maintenance of intensive animal production is driven by the greed and lack of care of the producers; and (c) the way forward is either vegetarianism or a return to traditional, extensive, small-scale, multipurpose farming.

Regarding (a) the FAO paper asserts that intensive production systems has some positive effects on animal welfare, and moreover that there is no reason to think that intensive production systems cannot be developed so as to provide high levels of

* Director of the Danish Centre for Bioethics and Risk Assessment and Professor of Bioethics at the University of Copenhagen

[1] FAO readings in ethics 2, Food and Agriculture Organization of the United Nations, Rome, 2005. Available as a pdf on the FAO website: http://www.fao.org/ethics/ser_en.htm.

P.B. Thompson (ed.) *The Ethics of Intensification*,

welfare. As for (b), the claim is that development of intensive animal production was mainly driven by economic factors beyond the control of individual farmers; and that for this reason, as Fraser puts it, "some of the standard claims made by critics of intensive animal production are seriously flawed" (p. 1). Finally, in connection with (c) Fraser argues that the way forward is to reform rather than abolish intensive animal agriculture.

It is timely that someone like Fraser, with a background in the study of animal welfare and ethics, is speaking out in favour of a more balanced discussion of intensive animal production. I agree with most of Fraser's conclusions; indeed I accept all of Fraser's main contentions except one. The exception is Fraser's generally wholesale rejection of the perspective of those urging greater emphasis on animal welfare and/or animal rights. Fraser's attitude to these (in effect) opponents is uncharitable and simplistic. Rather than saying that those who oppose intensive animal production are prompted to do so by a flawed understanding of the world, he should admit that there can be genuine, debatable, non-trivial disagreements about the underlying ethical principles here. It is also my view that the thinking leading Fraser to the conclusions he draws is occasionally questionable. In what follows these friendly criticisms of Fraser's influential essay will be presented.

The Link Between Intensive Production and Animal Welfare

Fraser argues compellingly that its critics often have a one-sided view of intensive animal production. For example, it is not true that intensive animal production will, in all cases, lead to farms being run by huge corporations. Thus in Denmark, which is the world's largest exporter of pork and a country where intensive livestock farming is the norm, the vast majority of pig farms are still family farms.

Furthermore, if animal welfare is defined in the way that it is most often defined by scientists studying animal welfare, the confinement systems typically used in intensive agriculture "appear to have both advantages and disadvantages" (p. 8). However, as Fraser himself notices, this crucial point in his line of argument very much depends on the way in which animal welfare is defined; and he is aware that many people would not accept the narrow definition of animal welfare he wishes to deploy here – a definition, that is, according to which welfare involves the absence of suffering and/or proper functioning. The FAO document mentions a Dutch study which found that "for some people, animal welfare depends on animals having freedom and living in natural environments" (p. 8). Recent Danish research in which I was involved made similar findings (Lassen et al. 2006).

Fraser's FAO paper seems to reject the idea that a wider definition of animal welfare should be taken seriously. Thus he says that according to such a definition of animal welfare "confinement systems are, by definition, incompatible with high animal welfare" (p. 8). In saying this he appears to assume that, in a discussion, serious moves cannot be made by ruling out opposing claims by definition. However, this cannot be true in general. Of course, if one simply closes a discussion by means of a definition made up just for the purpose of closing the discussion, this is not a serious move. But if a definition reflects a widely felt moral concern that is not patently mistaken, it cannot, it seems, be rejected out of hand.

Consider, to draw an analogy, a discussion about whether a person's welfare is eroded when another individual does something to that person without his or her prior consent. Here one could sincerely argue that according to one's definition of human welfare (i.e. one's preferred account of what constitutes a good human life) the answer is 'yes' – or, more guardedly, 'yes, in normal circumstances'. Would that not be a serious move? It seems that it would; and therefore a criticism of intensive agriculture based on a definition of animal welfare that includes natural living of some form should be taken seriously.

In this connection it is also worth pointing out that in Europe it has recently been decided, in the light of concern about animal welfare, that some of the most widely used confinement systems in animal production will be phased out. Thus in future pregnant sows in Europe will have to be loose-housed. Equally, egg producers will no longer be permitted to keep laying hens in traditional cages; the hens will have to be housed either in enriched cages or larger groups. But this certainly does not mean that intensive animal production will be phased out in Europe. It merely means that in the future, in Europe, intensive animal production will be based less on confinement of animals than it used to be.

It is interesting that some scientists (see, e.g., Savory 2004) criticize moves away from confinement systems in intensive animal production precisely because, as they see it, the animals (e.g. laying hens) are better off in confinement systems (battery cages) than they are in systems that are less confined (barn systems). Needless to say, these critics tend to deploy a narrow definition of animal welfare. There is, then, a real discussion to be had here – one that combines scientific perspectives and value issues. And this discussion is extremely relevant to public discussion about whether and how to reform intensive animal production so as to meet animal welfare concerns. Unfortunately, as a result of his rather polemical approach, Fraser simply stays out of this discussion. In doing so, he contrives to represent debate over animal welfare and intensive animal production more black and white than it really is.

Intriguingly, where the link between intensive animal production and animal welfare is concerned, the position taken by Fraser in the FAO document seems to share an important bias with those that he criticizes, for he in the paper focuses on the environment rather than on the animals themselves. This focus introduces a danger that the effects of a cornerstone in the development of intensive animal production – farm animal breeding – will be overlooked, or at least underestimated.

In parallel with the development of confinement systems and other production facilities that typify intensive farm animal production there has been an ongoing effort to breed farm animals so that, evermore efficiently, they produce meat, milk, eggs, and so on. Since the 1950s, average milk yields of dairy cattle in the developed countries have more than doubled. Similarly, the time it takes a broiler chicken to grow from hatching to readiness for slaughter has more than halved. It is true that some of this increase in efficiency is due to improved feeding, housing and disease control, but most of it derives from selective breeding.

Unfortunately, a negative side-effect of developments brought about by breeding has been a dramatic increase in various production-related diseases – notably, mastitis and metabolic diseases in dairy cattle and leg problems in broilers (Sandøe et al. 1999). There is very little that individual farmers can do about this. Particular in the poultry sector, animal breeding is controlled by a handful of multinational companies for whom the largest share of the market is found in regions where animal welfare has so far not been seriously recognized as an issue. In pig and cattle breeding centralisation is also increasing. Regrettably, it is difficult to see how legislation or market pressures could, in reality, prevent or significantly mitigate negative effects on animal welfare from farm animal breeding (Olsson et al. 2006).

In my view, then, selective breeding of farm animals for production is a significant part of the family of problems we now have with intensive agriculture. But sadly it has not received nearly as much attention as it deserves. This is partly because the critics of intensive production whom Fraser locks horns with in his paper do not seem always to have been aware of it. One serious weakness in the original FAO paper is that Fraser repeated the errors of those he criticizes on this important point.

Intensification Driven by Economic Forces – and So What?

In a section labelled "An alternative interpretation" Fraser attempted to put his finger on the forces that led to intensification of animal production in rich parts of the world from the 1950s onwards. Here he provides an economic explanation – one that he neatly summarises in the following way:

> In summary, this alternative hypothesis proposes that developments in the twentieth century, notably in transportation and food preservation methods, allowed a greatly expanded trade in animal products and consolidation of the food processing industry; that the resulting increase in competition caused periods when producers received very low profit per animal; and that these periods of low profit were a major factor contributing to the shift towards larger units and confinement housing, and also necessitated cost-cutting in factors such as space, staff time and other amenities. (p. 13)

Immediately after this passage, Fraser admitted that his economic explanation is "greatly over-simplified" (p. 13). He acknowledged that other factors such as a shortage of labour, rising wages and general developments in technology may have been at least as important as those he mentions.

However, in the case that Fraser tried to make against what he calls "the standard critique" of intensive animal production it does not really seem to matter which specific economic factors have driven developments. The key point is that these developments were not driven or directed by the voluntary choices of individual producers; they were determined by wider economic forces beyond the control of the producer. And they left the producer with just one choice: intensify your production or go out of business (and, of course, as we can now see, the vast majority of producers "chose" to go out of business).

Fraser eloquently summarized the way in which his explanation of the push for intensification is relevant to ethical discussion:

> In terms of production methods, it emphasizes not the macro-level features of increased farm size and use of confinement systems, whose influence on animal welfare is arguably mixed, but the micro-level features, specifically the cost-cutting required of producers at the same time that animal production intensified. In terms of economics, it suggests that the problem has not been excessive profit-taking by large corporations, but low and unpredictable profits and the constraints these place on producers. In terms of values and ethics, it suggests that the key problem is not the erosion of animal care values by producers as much as the values of consumers, expressed through their purchasing habbits, which leave producers little room for discretion in applying the animal care values that they may hold. (pp. 13–14)

A serious concern about this line of argument is that is it not at all clear why proponents of "the standard critique" should feel that they have been hit by it. Why would they want to deny that developments in animal production are driven by economic forces? Looking into the, in fact very short, section of his paper in which Fraser described the views of his opponents, one searches in vain for a quotation from anyone suggesting that the intensification was brought about by evil and greedy farmers.

What *does* emerge from the quotations in the relevant section here is a change in the values informing the main drivers of the business – roughly speaking, from care and good stockmanship to profit-seeking at all costs. But in these quotations, no claims are made concerning what is supposed to be the cause and what is supposed to be the effect in the relevant developments; and in fact there seems to be little or nothing in Fraser's account of the growth of intensification with which more radical critics of intensive animal production need to disagree.

So what really seems to divide Fraser and the adherents of the standard critique of intensive animal production is disagreement of one kind or another about values and strategies. The assumptions about values (V_1–V_3) and strategy (S) that underlie the position advanced in the original FAO ethics paper advocate for a reformist attitude to intensive agriculture, and they appear to be these:

V_1: *Loyalty to farmers and the farming community.*
V_2: *Animal welfare.* We should do our best to minimize stress, pain and other forms of suffering in animals we deal with.
V_3: *Consequentialism.* We should focus on how our actions will help to bring about the largest total good (or absence of bad things) rather than on what, in a more narrow sense, we ourselves do.
S: More good, in terms of farm animal welfare, will be brought about if people who care about animals embark on reforming intensive animal production than would brought about by such people spending their energy on promoting vegetarianism and/or small-scale extensive farming.

To me, it seems that each of these assumptions except V_2 might well be sincerely disputed by at least some of the critics of intensive animal production.

In his endorsement of V_1 Fraser, like me, is probably very much influenced by what he describes as "veneration of the farmer and farm family living in harmonious relationship with the land" (p. 4). Although relationships have become, even by Fraser's standards, much less harmonious with the development of intensive animal production, part of the old value framework still lives on in the spirit of the farming community. However, it is not clear why anyone who has not, through his or her background or professional experience, had the same contact with the farming community should share this loyalty.

Regarding V_3, clearly some of the critics with whom Fraser is engaging – notably Peter Singer and other utilitarians – are committed to a consequentialist approach. However, other critics are inspired by alternative theoretical standpoints such as the animal rights approach advocated by Tom Regan (Regan 2004). These critics may not accept consequentialist conclusions. They may argue that since animals have a right not to be treated in the way they are treated within intensive animal production systems, each of us has a duty to become a vegetarian, and that this remains so even if a different strategy might lead to less grave effects on animal welfare or to a smaller number of violations of animal rights.

Discussions within both academic philosophy and applied ethics demonstrate that the debate for and against consequentialism is not an easy one; there are knowledgeable, clever people on both sides of the discussion. Hence consequentialism cannot just be taken for granted in ethical discussion: it has to be argued for. The burden of the present observations, then, is not to criticize Fraser for being a consequentialist. It is to criticize him for assuming that nobody in their right mind would disagree with him on this point.

Turning to strategy, S is clearly a controversial assumption. Critics of intensive farming, notably Peter Singer (e.g. Singer 1993), have argued that the only effective strategy to reduce the suffering of animals in modern animal production is by promoting a vegetarian life-style. Thus Singer, it seems, would simply disagree with Fraser over the facts here. And since counterfactual claims about what would happen at the macro-level if one strategy rather than another is promoted are notoriously difficult to support empirically, Fraser really should not criticize Singer and other holding similar views in a dismissive way. More humbly, he should try to argue that, on balance, the available evidence speaks in favour of his strategy rather than Singer's.

Reformism – A Solution or a Cover-Up?

The last section of the original FAO Ethics Paper lists six "solutions that will promote animal welfare in a world where vast amounts of animal products continue to be eaten, and where much animal production will continue to follow intensive production models" (p. 15).

The first solution consists of finding "ways that animal care values can be encouraged and sustained" (p. 15). In my opinion this sounds highly programmatic and does not give us much guidance on what, specifically, to do. This impression is even stronger in the case of the fifth solution. Fraser set out this solution in the following way: "Fifth, as long-distance trade in animal products increases further through trade liberalization, we need to ensure that this trend does not lead to a new phase of near-zero profits and further constraints on producers' ability to act in ways that favour animal welfare" (pp. 16–17).

Fortunately, the four remaining suggestions are much more specific:

The second is to create economic incentives for producers to look after animal welfare. There are four examples of such incentives:

> (1) product-differentiation programs that provide premium prices for products produced according to specific standards; (2) government programs to help producers adjust to animal welfare standards, perhaps modelled after monetary incentives used to encourage conversion to organic methods; (3) purchasing agreements whereby corporate customers (chain restaurants, retail chains) agree to pay higher prices in return for guarantees of animal welfare standards; and (4) supply management programs that ensure that prices paid to producers reflect the cost of producing animal products in a manner that conforms to agreed animal welfare standards. (p. 16)

Fraser himself acknowledged that international competition and the lack of harmonisation across countries may undermine such incentives. In my view, only (1) and (3) are realistic in today's climate of free trade. They will allow animal welfare friendly production for segments – probably rather small segments – of the market. However, general government subsidies will run into trouble with international trade agreements; and general supply management programmes will face problems with competition from companies who do not run such programmes. Thus, in the foreseeable future, economic incentives will probably only affect the high-end segment of the market.

The third solution Fraser offered focused on "identifying and correcting the key management factors affecting animal welfare in all systems" (p. 16). Here there can be room for collaboration between farmer's or producer's organisations, research institutions, advisors, veterinarians and moderate animal welfare organisations. To some extent there can be a shared interest, since better management may simultaneously improve production and welfare. But the limit of this approach is, of course, economic: devoting resources to improved management will only be economically viable up to a point. Beyond that, there will be the usual dilemma between animal welfare and the bottom line.

The fourth solution was to adopt standards enforced by international law. For example, already as mentioned earlier in this commentary, the EU has a number of directives defining minimum standards for the housing, transport and slaughter of farm animals; and attempts are now being made to impose minimum standards world wide through the World Organization for Animal Health (OIE). These efforts may have some effect. However, there are problems getting beyond what might be considered an 'absolute minimum' of animal welfare; and there may well be huge problems with compliance and enforcement.

The final solution suggested in the FAO Ethics Paper was to promote a new ideal of what constitutes a good animal producer, "one that emphasizes a high level of animal management skill, scientific knowledge, staff management ability, a professional ethic of animal care and an appreciation of the need to conform to standards"

(p. 17). Certainly, this would be an important element in efforts to further a new vision of animal welfare that goes hand in hand with intensive animal production.

However, it remains to be seen whether these welcome-looking initiatives will have a significant impact on the welfare of farm animals – or whether they will, in effect, benefit only a minority of animals produced in well-off countries where some consumers are willing to pay for animal welfare. It remains possible, then, that all these well-meant initiatives would, in essence, operate as a cover-up for a crueller reality where the majority of production animals, thanks to market forces and the indifference of most of the players in the business, are living and dying in conditions ranging from the inadequate to the truly appalling.

Like Fraser I think that reformism is the only way forward. However, I am painstakingly aware, first, that I may actually serve as a useful idiot in my public defence of the view that it is possible to reform intensive animal production so as to improve animal welfare; and secondly, that the radical critics which Fraser so vehemently opposes have an important role to play. These critics help to raise awareness and force the industry to take steps in the right direction, including steps towards alliances with moderate pragmatics like Fraser and myself.

Chapter 14
Farm Animal Welfare: A Systemic Challenge

Richard Bawden*

There are two questions that are central to talk of improvements in any particular human enterprise: Who decides what constitutes "better" in any specific context? And what particular criteria for betterment are privileged in that context? As animal scientist David Fraser illustrates in his chapter on the intensification of livestock production, even in the case of something as seemingly straightforward as the welfare of farm animals, to address these questions seriously is to enter into a world of complexity, uncertainty, contextuality, and inevitable controversy.

As he states it, Fraser's purpose here is not to fight the fight against what he refers to as the Standard Critique of intensive animal production. This, as he sees it, essentially argues either for the rejection of animal agriculture altogether and the universal adoption of a vegetarian diet or for a return to extensive, non-confinement forms of livestock production. Rather, in accepting the reality that intensive animal agriculture is an embedded aspect of modern life, he seeks improvements to entire systems of livestock production that reflect concerns for values of animal care which can be expressed in acceptable practices and indeed, social policy, to that end.

This systemic challenge, as he would be the first to acknowledge, is no mean task: Like any 'agri-food system', intensive livestock production is a multi-dimensional, multi-functional, multi-stakeholder enterprise which operates within multi-faceted ecological, social, political, economic, cultural, and technological environments. There are countless actors and agents involved both in this networked system itself and in its environment, with each having her or his own motivations, concerns, and priorities; and each behaving according to her or his own understanding, knowledge, beliefs, prejudices, and emotions. Under such complex circumstances, there will be a host of different perspectives on the constitution of betterment, while the criteria that are appropriate to each perspective will together span a very wide spectrum of human interests. The level of interest in, focus on, and commitment to the welfare of the animals themselves, will typically vary enormously across this spectrum from consumer to processor to input supplier to policy maker to producer to activist and beyond – as well as to those scientists, technologists and ethicists in

* Dean Emeritus of the Hawkesbury Agricultural College, now a component of the University of Western Sydney in New South Wales, Australia

P.B. Thompson (ed.) *The Ethics of Intensification*,

particular who are committed, also for one reason or another, to promulgating the importance of the welfare of animals to all within the whole systemic complex.

Fraser is keenly aware of the multi-dimensionality of all of this, and in particular of the need for the articulation of a basis for decision makers, whoever they are, to improve the quality of their decisions about welfare improvement by employing multi-dimensional criteria and integrated 'mental' processes. In a number of his previous publications (e.g., Fraser 1995, 1999), his appeal has been for a 'bridging' between what he refers to as the 'two cultures' – of welfare science on the one hand and animal ethics on the other. His basic argument in support of such a convergence between natural science and philosophy has been that neither empirical information nor ethical reflection by themselves are adequate for answering questions about our proper relationships to animals of other species. His logic does not compel workers in each of these fields to be equally conversant with both scientific and philosophical aspects of the welfare issue, nor does he suggest that the disagreements about our proper relationships with other species will be forever resolved through such an endeavor of intellectual synthesis. Rather, his point is that while the arguments between the different parties may well continue, the premises will be closer together as scientists come to better understand some of the philosophical foundations of animal welfare and philosophers refine their ethical theories through incorporation of more sophisticated understanding of scientific principles. Even where goodwill exists to further such an intellectual and moral integration, the task is far from an easy one under circumstances where, as Fraser emphasizes here, there is considerable disagreement about what is actually meant by animal welfare from within, as well as between, scientific and ethical perspectives. For instance, in an earlier submission, Fraser, with his colleagues, distinguished between three approaches to animal welfare: (a) the natural living approach (where welfare is a function of the degree to which an animal lives a life that is judged to be as natural as possible), (b) the biological functioning approach (which places particular emphasis on physiological and behavioral functioning), and (c) the subjective experience approach (where the central concern is for the feelings of the animal) (Fraser et al. 1997). There is also considerable disagreement about the role of values in all science, including agricultural science, which, by convention, privileges the empirical at the exclusion of the normative (Alrøe and Kristensen 2002).

While he does not explicitly address the issue here, Fraser's call is not simply for attention to be focused solely on normative aspects of improvements in systems of intensive livestock production that reflect such values, but also on improvements in the way that those values are explored in synthesis with empirical knowledge about the welfare of animals generated through relevant scientific investigation. At base, he is thus as much concerned with the development of 'knowing systems' and their application in human judgment as he is with improvements to production systems that express that better knowing.

In this current piece, Fraser illustrates the further complexities of the matter that become immediately apparent when economic considerations are added to this mix and the focal issues are extended, beyond the twin domains of natural science and philosophy, out into the messiness of the economic realities of agri-food systems.

His intention here is not to investigate the conceptual significance and challenges including the social sciences into his 'bridging' exercise – thus extending his 'two cultures' into 'three' by embracing what Roling (2000) refers to as *gamma* as well as *beta* science – but of setting practical economic considerations as a conditioning factor of decision-making within the context of animal care.

Faced with the downward trends of declining profitability that have characterized their intensive animal production enterprises over recent decades, producers in family and factory farms alike, are under continual pressure to cut back on their costs. This in turn, Fraser argues, threatens the provision of the resources and the quality of the conditions that are essential to good welfare practices. There is little new in the observation that intensive animal producers, like many other farmers, are trapped within the notorious cost/price vise where the rate of increase in the costs of their inputs typically outstrips the rate of increase of the revenues that they get from the sale of their outputs. What is novel here is an appreciation of the impact that the nature of this 'farm problem' can have on animal care within intensive animal enterprises where, Fraser suggests, the attitudes of the producers towards their animals range from the 'callous to the caring' – as they probably always have. For those who are on the caring end of that spectrum, who continue to espouse those values, there are serious circumstantial impediments to their ability to act on them.

The classical response to the all too pervasive circumstances of declining terms of trade in agriculture is to address the productivity of the farming system and to seek improvements in the efficiency by which inputs into that system (costs) are converted into outputs (returns) from it. This productivity improvement response is the essential driver of the intensification of livestock production. Conceptually there are only three ways of achieving this end: (a) to reduce inputs without impairing the level of outputs, (b) to reduce inputs at a rate that exceeds consequential reductions in the level of outputs, and (c) to increase inputs at a level that is exceeded by the rate of increase in outputs. When translated into economic terms, two further options become available in practice through external interventions that can be policy-driven as well as consumer-led. Farm costs can be 'artificially' lowered through external support and/or farm prices can be 'artificially' raised through a variety of market mechanisms including quality bonus payments or price premiums that reward producers for adopting particular practices that are considered by policy makers and/or consumers, to be 'better'. Technological innovation plays a number of very important roles in productivity enhancement, essentially wherever capital can be substituted for labor. Technology has thus not only been vital in providing the infrastructures that are central to the process of animal confinement but, in its many guises, has become such a feature of intensive animal production that it is almost impossible to think of management practices that do not have a technological foundation. This is indeed so ubiquitous that such agri-food systems have become expressions of the industrialization imperative of modernism which, as Thompson (2004) observes, essentially reflects a pervasive worldview characterized by an uncritical faith that science and technology are inherently progressive. He is led to conclude that this pervasive mentality contributes, along with policy and technology, to the bind in which intensive animal producers find themselves.

Fraser meanwhile, submits that producers need to be protected from the cost/price economic pressures that they face, citing several examples of how this might be achieved. His context here is a myth-busting investigation that sets out to rebut what he sees as the often ‘rhetorical condemnations’ and frequently unsubstantiated claims directed at intensive animal agriculture by those of the school of what he calls Standard Critique. Fraser proposes an hypothesis about the root causes of intensification which is alternative to the basic arguments of this school of critics – which are couched so often, as he illustrates, in the colorful rhetoric of language like ‘corporate greed’ and ‘factory farming’ and ‘inhumane technologies’ – which he then uses as a foundation for articulating a different approach to welfare issues in intensive animal production systems.

In essence, the argument here is that the intensification of animal production is less a function of heartless corporatization driven by a greed for profit, than it is an expression of the responses from traditional family-owned enterprises to the economic forces that confront them. Accordingly, from an ethical perspective, the key problem with the intensification of livestock enterprises, at least as Fraser sees it, lies with values espoused by consumers, as expressed through the market, and not with the erosion of animal care values by the producers: It’s just that, in practice, they cannot afford to apply the animal care values that they may well hold, if they are to financially survive let alone thrive in the economic climate in which they are forced by the market place, to operate.

The solution to this problem, Fraser concludes, is to find ways to encourage and sustain strong animal care values, particularly through an emphasis on the technical, managerial, and ethical professionalism of producers working within well-designed and well-managed intensive production systems. This would need to be reinforced, he further argues, by the external implementation of economic support strategies and market interventions that would protect those producers from the cost-cutting pressures of the market place in such a manner that they could afford to express those values in their practices.

Logical though these arguments may be, they mask the full complexity of an issue that is truly systemic. It’s not just the consequences of the cost/price squeeze that threatens the integrity of livestock producers for ethical concerns within agri-food systems are not confined solely to the welfare of animals and nor are they the exclusive concern of producers. They include the potential threats that animal products represent to human health, as well as the potential negative impacts that pollutants from the enterprises represent to both the bio-physical and socio-cultural environments with which they are systemically inter-connected. There are too the considerable emergent issues that the introduction of the use of the so-called rDNA bio-technologies and other non-traditional genetic manipulations into intensive animal industries pose. And all the while, the range of outputs from intensive animal production systems, is being extended beyond food products to embrace what are often called ‘farmaceuticals’ – bio-synthetic drugs, vaccines, xeno-grafts and so on. These all are matters of concern to the whole citizenry and just as they share the benefits from intensive animal production so too must citizens bear some of the responsibilities for the production practices that are adopted. In other words, these ethical matters are as much social issues as they are issues for the producers themselves.

All of these issues must therefore be set within a context of their role as factors in the decision-making processes not just of producers themselves but of everyone else who ought to be involved in multi-criteria judgments what constitute improvements to the nature and dynamics of agri-food systems.

The challenge, then, is not just the design of more humane systems of agricultural production or more ethically-oriented systems of management or broader conceptions of the productivity of systems beyond an economic metric. The more fundamental issue is the need for profound attention to be paid to systems of knowing that underpin the prevailing mentalities and pervasive worldviews of all of those involved in the whole agri-food system including its livestock production subsystems – and this of course includes consumers as well as producers, and all others in between! The call really ought to be for attention to be paid to cognitive and spiritual systems of intellectual and moral processing that, through their further development and universal promulgation, could provide improved conceptual frameworks for more effectively exploring such complexity while also allowing a much fuller expression of the human potential for judgment making.

In this view, it is unfortunate that Fraser has not sought to explore the synthesis of the social sciences, and their associated 'culture' into his conceptual integration of the natural science culture with that of philosophy. As Roling (2000) emphasizes, the *Gamma* (social) sciences are basically about cognition; about understanding the way by which people collectively make sense out of the world about them as the basis for their judgments about, and subsequent actions in and to it. We need go no further than to refer back to the economics that Fraser introduces in his chapter here, for it is the neo-liberalist mantra of 'the greatest good for the greatest number' that surely represents the most pervasive worldview perspective that people in industrialized nations collectively use as the intellectual and moral framework for making their judgments and reaching their decisions. Care for animals is an externality in this context. Neo-liberalism is also entirely congruent with scientism as the other major ideological current of modernism; each fuels the other, most especially under the cultural circumstances that prevail where critical reflection is all too often afforded such a low pedagogical status within education.

We live in a world in which the unfettered and fast-globalizing market economy of neo-liberalism is the dominant ideology and where, as a consequence, the need for productivity gain in agricultural enterprises through innovation is pervasive in the face of the ever-declining terms of trade that farmers continue to experience. This is a world, moreover, in which climate change and terrorism and weapons of mass destruction and stem cell research and the price of gasoline claim the daily headlines away from the farm problem; and yet it is a world in which billions of people go to bed each evening ill with hunger and/or malnutrition because they have such limited capacity to meet even their most basic dietary needs.

This is also a world in which technical rationality of modernism has come to reign supreme and where the laboratories of techno-science have assumed the global status of a leviathan; yet it is a world where agriculture is increasingly devalued as a domain of serious intellectual and moral endeavor and where farming is discounted as an enviable social aspiration. It is a world in which the communicative void between the techno-scientific and economic rationalist experts on the one hand, and the citizens on the other, grows ever-wider and ever-deeper by the minute. The institutions of education and of research have so fragmented knowledge in this world, that even scholars from different disciplines – separated by epistemological as well as linguistic distinctions – are no longer able to effectively understand each other, let alone successfully collaborate in the exploration of the multi-faceted problematique that contemporary agri-food systems poses.

The work of David Fraser and his ilk in 'bridging the cultures' between animal welfare science and animal ethics is invaluable in raising consciousness and competencies among natural (*beta*) scientists and bio-ethicists alike. And this is finding valuable expression in much improved design and management of intensive animal production systems. Yet if the systemic complexity of the animal care issue is to be addressed in all of its multi-dimensionality, attention needs to be urgently paid to the inclusion of the social (*gamma*) sciences into the 'cultural bridging' project to which Fraser is clearly committed.

The challenge is fundamentally an epistemic one, for even the ability to think in terms of the 'systems' and their 'environments' that are central to the issue of dealing with the complexities of intensive animal production demands particular cognitive capabilities that need to be recognized and fostered (Bawden 2000).

Chapter 15
Ethics in Agricultural Change: Questions and Proposals for Development Processes

Andrew Dorward*

Introduction

Ethical issues associated with agricultural change have been the subject of a large literature appraising and evaluating agricultural investment and development projects, the green revolution and the development and regulation of bio-technology. Paradoxically, however, formal systems of ethical analysis for considering agricultural change have received relatively little explicit attention, a notable exception being the Nuffield bioethics reports. This is an unfortunate gap as there are major ethical questions that development policy makers and practitioners, and wider society, need to address. Nevertheless, it is also true that development policy makers and practitioners, not to mention wider society, work in an environment where identifying and then executing the ethically appropriate actions must be done in a timely fashion. The usefulness of ethical analysis thus requires a clear understanding of when and where ethical considerations come into play in the decision making process.

These issues need to be addressed, starting from the initial question of what the major ethical issues are in the development, dissemination and up take of technical change in agriculture. The first Ethics Paper in the FAO series was intended to address this question in a fashion that did not presuppose a great deal of formal training in ethical analysis. The identification of major ethical issues leads to questions about how these ethical issues should be considered and influence decisions about technical change, and who should be involved in such decisions, and how. Failure to explicitly address these questions means that political debates about agricultural change are easily bogged down in discussion at cross purposes, and the interests, rights, and values of different interest groups are treated inadequately and inconsistently. Policy processes then need to not only explicitly consider the ethical implications of particular policies, but also develop specific systems for addressing these issues and including them in normal processes of policy development, adoption and implementation.

* Professor of Development Economics, Department of Financial and Management Studies, School of Oriental and African Studies, University of London

P.B. Thompson (ed.) *The Ethics of Intensification*,

This volume has sought to stimulate wider consideration and debate of these issues, especially as they involve the main traditions of thought and analysis in philosophical ethics. Starting from the introduction by Thompson the various papers give a basic introduction to different ethical systems that can be applied in analysis, and consider a range of issues involved in the application of ethics to considering processes of agricultural change. What are the key policy lessons that emerge from this debate? After a very brief summary of salient points about different ethical systems and values as they relate to agricultural development, this chapter identifies critical elements of agricultural change that are likely to be most relevant to ethical analysis. This model of agricultural change then provides a platform for considering the practical questions that lie behind this book: how can ethical questions be helpfully brought to bear in agricultural development policy analysis and implementation?

Ethical Systems and Values in Development

Formal analysis is needed to understand often informal ethical systems applied by different people to evaluation of an agricultural change. The papers in this volume highlight the many different dimensions that there are in ethical systems and values. The FAO ethics paper on intensification lays out three systematic frameworks: costs and benefits, rights and duties, and virtues. Acknowledging at least these three general ways to approach ethical questions is particularly important for opening up debate over ethical goals and side constraints because there has been a tendency to adopt technical approaches to development that become married to one mode of ethical thinking in a somewhat uncritical manner. This may happen because analysts have become complacently accepting of a particular ethical framework, but it may also happen when a technical analysis discloses objective information that will simply be more useful in one ethical framework than another. Thus it may often be the case that studies from the agricultural and social sciences "bias" ethical considerations in favour of a cost-benefit or utilitarian approach not because scientists explicitly hold those values, but simply because the predictive application of technical models can be easily incorporated into an ethical framework that valorises costs and benefits.

Discussion throughout the collection of papers illustrates the variety of social, religious and other bases of ethical values and the variety of systems and process of change in ethical values and systems. Of particular relevance to agricultural development and intensification are the differing (and variously conflicting and complementary) emphases on processes and outcomes; on social relations, production and consumption, and environmental values; on natural versus human processes and commodities; on precaution versus progress; and on sustainability of inputs, process, and outputs. Each of these dimensions of evaluation of development are fundamental to the different ways that different groups view development, and they reiterate the importance of recognising development as contested process where power relations play a major role in defining development goals and determining 'development' paths and outcomes. Development then becomes a political process where institutional and technical change interact across multiple dimensions of interest and activity and across different scales (from global down to community and family). Some of the debates and issues here are illustrated by the emergence of the millennium development goals (and particularly of poverty reduction) in international development policy alongside increasing emphasis on rights based approaches to development, the trade-offs and complementarities between and among these, and their differing implications for north-south relations.

How then can ethical considerations contribute in practical ways to the complex and contested processes of agricultural development? There is no simple answer to this, but the immediate contribution of this volume is to draw attention to the ethical considerations which underlie many of the conflicts around agricultural development and to relate these to options and processes in agricultural change. The previous chapters thus represent initial steps that need to be taken further in two ways; first the various ethical, technical, institutional, social and other dimensions of the processes and outcomes of agricultural development need to be related to each other in a wider analytical framework; this analysis itself then needs to be related to the question of agency. This should then lead to elucidation of who plays what roles in agricultural development processes and how these relate to ethical considerations – or how ethical considerations can be helpfully brought to bear on these roles. In this chapter we therefore put forward first a conceptual framework relating ethical, technical, institutional, social and other dimensions of the processes and outcomes of agricultural development. We then suggest how this framework can guide practical approaches for ethical appraisal in policy, program and project appraisal.

Processes of Agricultural Change

Figure 15.1 presents a schematic framework linking the range of major issues relevant to ethical considerations of agricultural change. On the left hand side of the diagram we identify the characteristics of technical and institutional innovations as a major focus of attention. Emphasis is given to both technical and institutional innovation as fundamental elements of agricultural development, and these innovations need to be assessed in terms of the nature and extent of the impact of their adoption to in turn allow them to be characterized in ways that draw out the ethical implications of such adoption.

Change and innovation may be helpfully classified and viewed here from a number of angles. We may want to consider the nature of change in terms of the balance between technical and institutional 'instruments' of change and the emphasis on social, economic, cultural outcomes. Alternatively or additionally we might consider the scope of change and its nature, scale and boundaries and relate these to farm, community, district, national, global analysis; to upstream and downstream relations in ecosystems or commodity chains; or to property rights, food, or labour

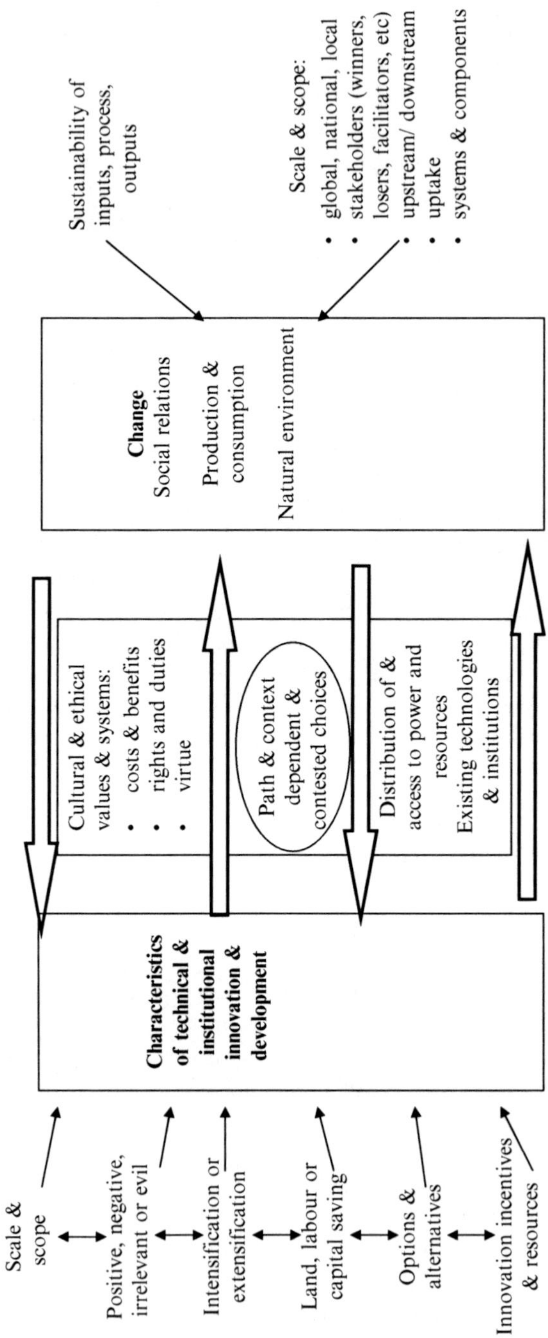

Fig. 15.1 Agricultural development processes

systems. These may be related to different processes of change in (for example) innovation systems (affecting resources, relations, incentives), power relations, or the distribution of resources. In laying stress on the multiplicity of ways that change processes can be understood, we can make a break with a mode of development thinking which sees change processes in purely objective technical terms, and which finds ethical significance only in the outcome of change processes.

In drawing out the ethical implications of change a number of papers in this volume have discussed the different social, economic, ecological and other effects of different types of intensification and extensification in different contexts, and Camacho has introduced the concept of technologies which may be in different ways wrong, evil, positive, negative, or irrelevant: this concept may be equally applicable to institutions. The nature of intensification and extensification and the classification of technologies (and institutions) however in turn depend upon innovation characteristics such as scale and scope (determining for example the nature and extent of impact on different people and ecosystems), whether they are labour saving or land saving (with differing effects on labour demand, wages, employment opportunities, productivity, competitiveness and incomes for poor and non-poor groups), the options and alternatives to these innovations, and the incentives and resources for their adoption.

These innovation characteristics and their ethical implications are then highly context dependent in two ways.

First, current technical, institutional, economic, social, political and ecological conditions will determine the particular effects of the introduction of an innovation on different people and agro-ecosystems and consequently the characteristics of that innovation that are important within that context. In other words, understanding how actors themselves understand ethical norms, responsibilities and desiderata will improve our ability to understand and predict the development process. This represents a break from social science models that interpret the entire range of human action through the lens of self-interested optimization. This is shown in the lower part of the central box in Fig. 15.1, which also makes the point that context dependency is also path dependency, as context depends upon previous changes (on the right hand side of Fig. 15.1), which themselves arose from the interaction of earlier innovation characteristics and their context, and contested choices about the adoption and rejection of different innovations.

Context and path dependency also arise, however, as a result of the evolution of cultural and ethical value systems. Just like other social and material conditions discussed above, these affect the perceptions of innovations, the contests about adoption (and adaptation) choices, and hence wider processes of change. They are also, in turn, affected by these processes of change, with evolving and diverse (and of course contested) emphases on and conceptualizations of costs and benefits, rights and duties, and virtue. It is not simply a matter of recognizing that some actors may be more focused on rights or virtues than on what they gain or lose in a development process, considerations reflecting contrasting or contradictory ethical orientations also engage one another through political debate, on the one hand, and through processes of imposition and resistance, on the other. The mutual engagement of contradictory ethical values through processes of verbal, political or even violent struggle introduces dynamism – a change process – into ethical commitments themselves.

Bringing Ethics into Agricultural Development Praxis

How does the brief discussion in the previous two sections of ethical systems and development processes help as we consider how to usefully bring ethics into agricultural development praxis?

The framework of Fig. 15.1 suggests that development agents and policy makers who seek to influence the direction, processes and outcomes of agricultural development face significant ethical responsibilities in this task. Such actors have always understood that development is an inherently normative idea, and that a successful development process is "ethical" at least in the sense that its outcomes represent progress. Figure 15.1, along with the full complement of essays in this book, shows that there are multiple ways in which success can be measured. As such, development practice can neither escape the possibility of succeeding on one scale while failing on another, nor dodge the need to craft policies and methods that reflect multiple standards and approaches to the extent possible.

The discussion in the second section suggests that the conflicts that arise in development processes might be more easily understood and addressed if the ethical systems that underlie them were made more explicit. Here, the scientific and technical training of development specialists may have failed them. In attempting to remain neutral or "objective", practitioners have made decisions that appear to lack any clear ethical rationale. In fact, the implicit utilitarian underpinnings of agricultural development policies provide perfectly legitimate ethical bases for policies and strategies designed to improve food availability, rural incomes and nutrition. Owning up to these underpinnings would facilitate a more honest and straightforward conversation and negotiation in cases where these policies or strategies are resisted because they are perceived to threaten either the rights of affected parties, or traditional practices seen as critical to the preservation of group identity or personal virtue.

How then should development agents and policy makers address their ethical responsibilities, and how can they develop a practical understanding of ethical issues as they work with and for the many different stakeholders in agricultural development? These are complex, multi-level and ultimately political and constitutional questions, and there will be no simple and universal answers, indeed the discussion in this chapter stresses the pervasiveness of conflict, context and path dependency in both ethical systems and development. However it is possible (and the argument here is that practitioners ought) to initiate debate on specific policies or strategies by proposing a process of ethical appraisal that may (or should) be applied in policy, programme or project appraisal to guide policy, programme and project planning and implementation. Such a process might involve three steps: development process analysis, stakeholder analysis, and ethical analysis. Each of these steps is described briefly in turn.

First an analysis of the processes involved in particular agricultural development situations is needed to identify the major options and issues in question. This analysis could use the framework outlined above and in Fig. 15.1 to identify the major characteristics of the innovations under consideration, in the context of cultural and ethical values and systems on the one hand and of current technical, institutional, economic, social, political and ecological conditions on the other (the latter involving the distribution of and access to power and resources and existing technologies and institutions, sustainability, social relations, production, consumption and the natural environment as outlined in Fig. 15.1).

Following from this process analysis would be an analysis of the stakeholders with interests in this process. Stakeholder analysis is a widely used technique in development planning and appraisal, involving identification of the stakeholders and description and analysis of characteristics, interests, rights, responsibilities and interests relevant to the development process and changes under consideration. This would need to relate back to the innovation characteristics and contexts examined in the prior process analysis, and to both the ethical values and systems of stakeholders and to their position within that ethical system (for example as people with particular rights, responsibilities, or vulnerabilities).

The third stage in ethical appraisal would be a formal analysis of the ethical issues arising from the process and stakeholder analyses. Initially this might involve a formal attempt at an objective analysis of relevant (and perhaps competing) ethical systems and the issues arising from proposed or emerging institutional and technical change. There might be considerable value here in the development and use of some sort of 'primer for ethical analysis in agricultural development', which might contain an overview of development issues and ethical systems (drawing, for example, on material in this collection), descriptions and sources on analytical methods, and checklists of issues. More thorough ethical appraisal would have to engage with issues of subsidiarity not just in development processes but in ethical appraisal itself, and would need to engage stakeholders in consideration of these issues and in negotiations about their ethical systems and innovation choices. Such engagement would then continue through and be integrated with other appraisal, planning, implementation, monitoring and evaluation activities – and indeed would need to be added to these with a light touch to prevent it from becoming over burdensome.

Conclusions

Ethical issues associated with agricultural change lie at the core of many rural development, environmental, food and agricultural policy debates, but are seldom used to expose the real causes of difference and conflict between protagonists in these debates. Similarly technical (natural science and socio-economic) methodologies for appraisal of agricultural development investments are founded upon particular ethical systems, but again these are seldom explicitly recognized or debated. As a result development debates are often unnecessarily confused while major ethical questions associated with agricultural development are not addressed in technical and political decisions.

This is not an acceptable state of affairs. For it to be addressed there is a need for (a) political will to bring these issues into mainstream discussions about agricultural development and (b) accessible tools and methodologies to enable agricultural development professionals (whether natural or social scientists) to bring ethical questions explicitly into the centre of development processes. Political will and the technical tools for implementing it need to be developed together. The papers in this collection provide valuable foundations for such developments, both demonstrating the importance of ethics in agricultural development and identifying the specific ethical issues that need to be addressed and their relationship with different kinds of agricultural development.

This concluding chapter has suggested in very broad terms a practical system for bringing ethical insights and appraisal into political and technical development processes, involving process, stakeholder and ethical analysis in normal agricultural development appraisal, planning, implementation, monitoring and evaluation activities. Having an explicit and clearly articulated statement of the ethical issues raised by any given policy proposal or development strategy would not necessarily put an end to conflict and struggle. Yet it might well avert some needless conflict and struggle arising from simple confusion, miscommunication and misunderstanding about what really matters to stakeholders (including policymakers themselves) in the development process.

Acknowledgments This paper has benefited substantially from Paul Thompson's editorial amendments, which are gratefully acknowledged.

Participants at the Conference in January 2004 at the Wye Campus of Imperial College London

Simon Anderson, Imperial College London
David Andow, University of Minnesota (US)
Deryke Belshaw, University of East Anglia and Oxford Centre for Mission Studies (UK)
Lawrence Busch, Michigan State University
Luis Camacho, University of Costa Rica (Costa Rica)
Andrew Dorward, Imperial College London
Nigel Dower, University of Aberdeen
Steve Esquith, Michigan State University
Michiel Korthals, Wageningen University (NL)
Jonathan Kydd, Imperial College London
Michael Lipton, University of Sussex (UK)
John Mason, member of local rural community (UK)
Jamie Morrison, Imperial College London
Clare Palmer, Furness College Lancaster University
Rodrigo Paz, Imperial College London/Bolivia Student
John Perkins, The Evergreen State College
Colin Poulton, Imperial College London
Peter Sandøe, University of Copenhagen
Allan Schmid, Michigan State University
Clive Stannard, FAO Rome
Paul Thompson, Michigan State University
Jeff Waage, Imperial College London
Lowell Woodcock, University of Sussex

Bibliography

Adams AW, Craig JV (1985) Effect of crowding and cage shape on productivity and profitability of caged layers: a survey. Poultry Science 64: 238–242

AFAC (undated) Alberta farm animal care. Alberta Farm Animal Care Association, Calgary. http://www.afac.ab.ca/index.htm. Accessed February 2005

Alrøe HF, Kristensen ES (2002) Towards a systemic research methodology in agriculture: rethinking the role of values in science. Agriculture and Human Values 19: 3–23

Amado J (1974) Gabriela, clove and cinnamon. Avon/Bard Books, New York

Aman K (ed) (1991) Ethical principles for development: needs, capacities or rights. Proceedings of the IDEA/Montclair Conference. Montclair College: The Institute for Critical Thinking, Upper Montclair, NJ

Amin S (1976) Unequal Development: An Essay on the Social Formations of Peripheral Capitalism. Pearce B (trans). Monthly Review Press, New York

Amin S (1977) Imperialism and unequal development. Translated from the French. Harvester Press, Hassocks, England

Animal Place (undated) Why even care about farmed animals? Animal Place, Vacaville, CA. http://www.animalplace.org/why.html. Accessed January 2005

Anonymous (1989) How Astrid Lindgren achieved enactment of the 1988 law protecting farm animals in Sweden. Animal Welfare Institute, Washington, DC

Appleby MC (1999) What should we do about animal welfare? Blackwell Science, Oxford

Balfour Lady EB (1950) The living soil: evidence of the importance to human health of soil vitality, with special reference to national planning. Devin-Adair, New York

Barnard CJ, Hurst JL (1996) Welfare by design: the natural selection of welfare criteria. Animal Welfare 5: 405–433

Bawden R (1990) Of agricultural systems and systems agriculture: systems methodologies in agricultural education. In: Jones JJ, Street P (eds) Systems theory applied to agriculture and the food chain. Elsevier, London, pp 305–325

Bawden RJ (2000) Valuing the episteme. Cybernetics and Human Knowing 7: 5–25

Bawden RJ, Ison RL (1992) The purposes of field crop ecosystems: social and economic aspects. In: Pearson C (ed) Field crop ecosystems. Elsevier, London, pp 11–35

BBC Radio 4, 4 December 2003, Costing the Earth

Beaudreau BC (1999) Energy and the rise and fall of political economy. Greenwood Press, Westport, CT

Bell D, Chase B, Douglass A, Hester P, Mench J, Newberry R, Shea-Moore M, Stanker L, Swanson J, Armstrong J (2004) UEP uses scientific approach in its establishment of welfare guidelines. Feedstuffs 76: 1–9

Benjamin M (1990) Splitting the difference: compromise and integrity in ethics and politics. University Press of Kansas, Lawrence, KS

Berlin I (1979) Four essays on liberty. Oxford University Press, Oxford

Berlin I (1992) The pursuit of the ideal. In: Hardy H (ed) The crooked timber of humanity. Vintage, New York, pp 1–2

Bhardwaj M, Maekawa F, Niimura Y, Macer DRJ (2003) Ethics in food and agriculture: views from FAO. International Journal of Food Science & Technology 38 (5): 565–577

Biffen RH (1905) Mendel's laws of inheritance and wheat breeding. Journal of Agricultural Science 1: 4–48

Bizup J (2003) Manufacturing culture: vindications of early Victorian industry. University of Virginia Press, Charlottesville, VA

Bogaard A (2005) 'Garden agriculture' and the nature of early farming in Europe and the Near East. World Archeology 37 (2): 177–196

Boserup E (1966) The conditions of agricultural growth; the economics of agrarian change under population pressure. Aldine, Chicago, IL

Brady E (2003) The aesthetics of the natural environment. Edinburgh University Press, Edinburgh, UK

Brady E (2006) The aesthetics of agricultural landscapes and the relationship between humans and nature. Ethics, Place and Environment 9 (1): 1–19

Brambell FWR (chairman) (1965) Report of the Technical Committee to enquire into the welfare of animals kept under intensive livestock husbandry systems. Her Majesty's Stationery Office, London

Broom, DM (1991) Animal welfare: concepts and measurement. Journal of Animal Science 69: 4167–4175

Buchanan A (2006) Taking the human out of human rights. In: Martin R, Reidy DA (eds) Rawls's law of people's. Blackwell, Malden, MA, p 153

Burger-Fischer v. Degussa (2000) 65 F. Supp.2d 248; Agreement between the Government of the United States of America and the Government of the Federal Republic of Germany concerning the Foundation "Remembrance, Responsibility, and the Future." Berlin, Germany, 17 July 2000

Burkhardt J (1997) The inevitability of animal biotechnology? Ethics and the scientific attitude. In Holland A, Johnson A (eds) Animal biotechnology and ethics. Chapman & Hall, London, pp 114–142

Burkhardt J (1999) Scientific values and moral education in the teaching of science. Perspectives on Science 7 (1): 87–110Busch L (2000) The moral economy of grades and standards. Journal of Rural Studies 16: 273–283

Busch L (2002) The implications of global standards for national agricultural research. In: Bigman D (ed) Globalization and the developing countries: emerging strategies for rural development and poverty alleviation. CAB International, The Hague, pp 171–184

Busch L, Bain C (2004) New! Improved? The transformation of the global agrifood system. Rural Sociology 69: 321–346

Camacho L (1989) On technology and values. In: McLean GF, Pegoraro O (eds) The social context and values, perspectives of the Americas. University Press of America, Lanham, MD, pp 125–139

Camacho L (1993) Ciencia y tecnología en el subdesarrollo. Editorial Tecnológica de Costa Rica, Cartago

Carlson A (2000) Aesthetics and the environment: the appreciation of nature, art and architecture. Routledge, New York/London

CEC (2006) Communication from the Commission to the European Parliament and the Council on a community action plan on the protection and welfare of animals 2006–2010. Commission of the European Communities (CEC), Brussels. http://europa.eu.int/comm/food/animal/welfare/com_action_plan230106_en.pdf. Accessed April 2006

Cherfas J (1996) Sustainable food systems. In: Mepham B (ed) Food ethics. Routledge, London, pp 35–48

CIWF (Compassion in World Farming) Trust (2003) The welfare of broiler chickens in the European Union. Compassion in World Farming Trust, London

Cleaver HM Jr. (1972) Contradictions of the Green Revolution. American Economic Review 62 (½): 179

Coats CD (1989) Old MacDonald's factory farm. Continuum Publishing, New York

Cochrane WW (1993) The development of American agriculture: a historical analysis. 2nd edn. University of Minnesota Press, Minneapolis, MN, pp 96–105
Cohen MN (1989) Health and the rise of civilization. Yale University Press, New Haven, CT
Cooper D (1998) Intervention, humility and animal integrity. In: Holland A, Johnson A (eds) Animal biotechnology and ethics. Kluwer, Amsterdam, pp 145–155
Cowan CW, Watson PJ (eds) (1992) The origins of agriculture, an international perspective. Smithsonian Institution Press, Washington, DC
Crocker DA (1991) Insiders and outsiders in international development ethics. Ethics and International Affairs 5: 149–174
Crocker DA (1998) Florecimiento humano y desarrollo internacional: la nueva ética de capacidades humanas. Editorial de la Universidad de Costa Rica, San José
Crocker DA, Linden T (eds) (1998) Ethics of consumption. Rowman & Littlefield, Lanham, MD
Curtin D (2005) Environmental ethics for a postcolonial world. The Rowman & Littlefield Publishers, Lanham, MD, p 21
Danmarks Statistik (2004) Landbrug 2003 (data updated 2 May 2003)
Dawkins MS (1988) Behavioural deprivation: A central problem in animal welfare. Applied Animal Behaviour Science 20: 209–225
Dawkins MS (1990) From an animal's point of view: motivation, fitness, and animal welfare. Behavioural and Brain Sciences 13: 1–9, 54–61
Dawkins MS (1998) Evolution and animal welfare. Quarterly Review of Biology 73: 305–328
Dolan EF Jr. (1986) Animal rights. Franklin Watts, New York/ Dordrecht, The Netherlands
Dower N (2005) The nature and scope of global ethics and the relevance of the Earth Charter. Journal of Global Ethics 1 (1): 25–43
Duncan IJH (1993) Welfare is to do with what animals feel. Journal of Agricultural and Environmental Ethics 6 (Supplement 2): 8–14
Egziabher T (2004) Nuffield Council should be more ethical on GM crops. Gaia Foundation, 12 January 2004
Eide A, Eide WB, Goonatilake S, Gussow J, Omawale (eds) (1984) Food as a human right. The United Nations University, Tokyo, p vi
Elliot R (1997) Faking nature. Routledge, London
Esquith S (2009) The Political Responsibilities of Everyday Bystanders. The Pennsylvania State University Press, University Park, PA
Eymard-Duvernay F (1995) La négociation de la qualité. In: Nicolas F, Valceschini E (eds) Agro-alimentaire: une economie de la qualité. INRA and Economica, Paris, pp 39–48
FAO (Food and Agriculture Organization of the United Nations) (1996) World Food Summit, 13–17 November 1996, Rome, Italy. Rome Declaration on World Food Security. World Food Summit Plan of Action Step 5. http://www.fao.org/docrep/003/w3613e/w3613e00.htm. Accessed 28 January 2008
FAO (Food and Agriculture Organization of the United Nations) (2000a) Priority area for interdisciplinary action on ethics in food and agriculture. http://www.fao.org/ethics/index_en.htm. Accessed 28 January 2008
FAO (Food and Agriculture Organization of the United Nations) (2000b) Reports of the panel of eminent experts on ethics in food and agriculture. http://www.fao.org/ethics/exp_en.htm. Accessed 28 January 2008
FAO (Food and Agriculture Organization of the United Nations) (2001a) Report of the panel of eminent experts on ethics in food and agriculture. First session 26–28 September 2000, Rome. http://www.fao.org/DOCREP/003/X9600E/X9600E00.HTM. Accessed 28 January 2008
FAO (Food and Agriculture Organization of the United Nations) (2001b) Ethical issues in food and agriculture. FAO Ethics Series, Rome. ftp://ftp.fao.org/docrep/fao/003/X9601e/X9601e00.pdf. Accessed 29 January 2008
FAO (Food and Agriculture Organization of the United Nations) (2001c) Genetically modified organisms, consumers, food safety and the environment. FAO Ethics Series, Rome. ftp://ftp.fao.org/docrep/fao/003/X9602e/X9602e00.pdf. Accessed 29 January 2008

FAO (Food and Agriculture Organization of the United Nations) (2003a) FAO expert consultation on food safety: science and ethics, 3–5 September 2002, Rome, Italy. ftp://ftp.fao.org/docrep/fao/006/j0776e/j0776e00.pdf. Accessed 29 January 2008

FAO (Food and Agriculture Organization of the United Nations) (2003b) Report of the panel of eminent experts on ethics in food and agriculture. Second session 18–20 March 2002, Rome. http://www.fao.org/DOCREP/005/Y8265E/Y8265E00.HTM. Accessed 28 January 2008

FAO (Food and Agriculture Organization of the United Nations) (2004a) Food balance sheet. FAOSTAT database: 27 August 2004 update, Rome. http://faostat.fao.org/faostat/form?collection=FBS&Domain=FBS&servlet=1&language=EN&hasbulk=1&hostname=faostat1&version=int&macroDomain=SUA%20data&macroData=Balance%20Sheet%20data. Accessed January 2005

FAO (Food and Agriculture Organization of the United Nations) (2004b) Food balance sheet. FAOSTAT database: 27 August 2004 update, Rome. http://faostat.external.fao.org/faostat/form?collection=FBS&Domain=FBS&servlet=1&hasbulk=0&version=ext&language=EN. Accessed January 2005

Food Ethics Council (2001) Farming animals for food: towards a moral menu. FEC, Southwell, UK

Fraser D (1995) Science, values and animal welfare: exploring the inextricable connection. Animal Welfare 4: 103–117

Fraser D (1999) Animal ethics and animal welfare science: bridging the two cultures. Applied Animal Behaviour Science 65: 171–189

Fraser D (2001a) The "new perception" of animal agriculture: legless cows, featherless chickens, and a need for genuine analysis. Journal of Animal Science 79: 634–641

Fraser D (2001b) Farm animal production: changing agriculture in a changing culture. Journal of Applied Animal Welfare Science 4: 175–190

Fraser D (2003) Report of the FAO's Panel of Eminent Experts on Ethics in Food and Agriculture. http://www.fao.org/DOCREP/005/Y8265E/y8265e00.htm

Fraser D (2005) Animal welfare and the intensification of animal production: an alternative interpretation. FAO Readings in Ethics No. 2. Food and Agriculture Organization of the United Nations, Rome

Fraser D (2006a) Caring for farm animals: pastoralist ideals in an industrialized world. In: Waldau P, Patton K (eds) A communion of subjects: animals in religion, science and ethics. Columbia University Press, New York, pp 547–555

Fraser D (2006b) Animal welfare assurance programs in food production: a framework for assessing the options. Animal Welfare 15: 93–104

Fraser D (2008a) Toward a global perspective on farm animal welfare. Applied Animal Behaviour Science (in press)

Fraser D (2008b) Understanding animal welfare: science in its cultural context. Blackwell, Oxford

Fraser D, Weary DM, Pajor EM, Milligan BN (1997) A scientific conception of animal welfare that reflects ethical concerns. Animal Welfare 6: 187–205

Fraser D, Mench J, Millman S (2001) Farm animals and their welfare in 2000. In: Salem DJ, Rowan AN (eds) State of the animals 2001. Humane Society Press, Washington, DC, pp 87–89

Freeman D (1974) Technology and society. Rand McNally, Chicago, IL

Gamborg C, Sandøe P (2003) Breeding and biotechnology in farm animals: ethical issues. In: Levin R, Reiss M (eds) Key issues in bioethics. Routledge, London, pp 133–142

García Márquez G (1971) One hundred years of solitude. Bard Avon, New York

Goddard NPW (1981) The Royal Agricultural Society of England and agricultural progress, 1838–1880. Ph.D. dissertation, University of Kent, Canterbury, pp 7–12, 74–98

Goodman D, Watts M (1997) Globalizing food: agrarian questions and global restructuring. Routledge, London

Goodpaster K (1978) On being morally considerable. Journal of Philosophy 75: 308–325

Goonatilake S (1984) Food culture and human rights. In: Eide A, Eide WB, Goonatilake S, Gussow J, Omawale (eds) Food as a human right. The United Nations University, Tokyo, pp 102–103

Goulet D (1965) Etica del desarrollo. Estela/IEPAL, Barcelona, Montevideo

Goulet D (1971) The cruel choice: a new concept in the theory of development. Atheneum, New York/University Press of America, Lanham, 1985

Goulet D (1989) Tareas y métodos en la ética del desarrollo. Revista de Filosofía de la Universidad de Costa Rica 66: 293–306

Goulet D (1995) Development ethics: a guide to theory and practice. The Apex Press, New York

Greenpeace. Golden Rice: all glitter, no gold. http://www.greenpeace.org/international/news/failures-of-golden-rice. Accessed 10 November 2007

Gussow J (1991) Chicken little, tomato sauce, and agriculture: who will produce tomorrow's food?. Bootstrap Press, New York

Habermas J (1998) The inclusion of the other: studies in political theory. Cronin C, De Greiff P (eds) MIT Press, Cambridge, MA

Harrison R (1964) Animal machines. Vincent Stuart, London

Harwood D (1928) Love for animals and how it developed in Great Britain. Republished (2002) (Preece R, Fraser D, eds). Dix Harwood's love for animals and how it developed in Great Britain. Edwin Mellen Press, Lewiston, ME

Held D (2003) From executive to cosmopolitan multilateralism. In: Held D, Koenig-Archibugi M (eds) Taming globalization: frontiers of governance. Polity, Malden, MA, pp 160–186

Hemsworth PH, Coleman GJ (1998) Human–livestock interactions: the stockperson and the productivity and welfare of intensively farmed animals. CAB International, Wallingford, UK

Oruka HO (1997) Practical philosophy in search of ethical minimum. East African Educational Publishers, Nairobi, pp 84–88

Hill T (1983) Ideals of human excellence and preserving natural environments. Environmental Ethics 5: 211–224

Holland A (1997) Species are dead! Long live genes! In: Holland A, Johnson A (eds) Animal biotechnology and ethics. Kluwer, Amsterdam, pp 225–240

Holland A, Johnson A (eds) (1997) Animal biotechnology and ethics. Kluwer, Amsterdam

Holmes S (1995) Passions and constraints; on the theory of liberal democracy. University of Chicago Press, Chicago, IL, pp 13–41

Howard Sir A (1940) An agricultural testament. Oxford University Press, London

Hull DL (1988) Science as a process: an evolutionary account of the social and conceptual development of science. University of Chicago Press, Chicago, IL

Human Rights Watch (2005) The dark side of Ethiopia's "green revolution." http//hrw.org/English/docs/2005

Humane Farming Association (undated) Factory farming. HFA, San Rafael, CA. http://www.hfa.org/factory/index.html. Accessed January 2005

Hunton P, Fairhurst G (2002) 2001 international egg costings show ongoing price drop. World Poultry 18 (1): 20–23. http://www.agriworld.nl/public/file/pdf/20060922-wp_18.1_k4_n1.pdf. Accessed October 2007

Inge MT (ed) (1969) Agrarianism in American literature. The Odyssey Press, New York

Jamison R (2003) Certified organic farming in Washington State: a feminist empiricist examination. MES thesis, The Evergreen State College, December 2003

Johnson GL (1976) Philosophic foundations: problems, knowledge and solutions. European Review of Agricultural Economics 3: 207–234

Johnson GL (1982) Agro-ethics: extension, research and teaching. Southern Journal of Agricultural Economics (July): 1–10

Kant I (1785) Was heisst Aufklärung.

Kathuria H, Giri J, Tyagi H, Tyagi AK (2007) Advances in transgenic rice biotechnology. Critical Reviews in Plant Sciences 26: 65–103

Kautsky K (1988) The agrarian question. Translated from the German by P. Burgess. Zwan, London

Kiley-Worthington M (1989) Ecological, ethological, and ethically sound environments for animals: toward symbiosis. Journal of Agricultural Ethics 2: 323–347

Kloppenburg R Jr. (2004) First the seed: the political economy of plant biotechnology, 1492–2000. University of Wisconsin in Press, Madison

Kolkman J (1987) Animal care: livestock and poultry on today's farm. Christian Farmers Federation of Alberta, Edmonton, Canada

Korthal, M (2004) Before dinner. Springer, Dordrecht, The Netherlands

Kriflik LS, Yeatman H (2005) Food scares and sustainability: a consumer perspective. Health, Risk & Society 7 (1): 11–24

Kuehn BM, Kahl SC (2005) The stall in sow housing. Journal of the American Veterinary Medical Association News (1 January 2005). http://www.avma.org/onlnews/javma/jan05/050101a.asp. Accessed January 2005

Kymlicka W (1995) Multicultural citizenship: a liberal theory of minority rights. Oxford University Press, Oxford

Ladrière J (1977) Les enjeux de la rationalité. UNESCO, Paris

Lane RE (1998) The road not taken: friendship, consumerism and happiness. In: Crocker DA, Linden T (eds) Ethics of consumption. Rowman & Littlefield Publishers, Lanham, MD, pp 218–248

Lassen J, Sandøe P, Forkman B (2006) Happy pigs are dirty! – conflicting perspectives on animal welfare. Livestock Science 103: 221–230

Lebret LJ (1959) Dynamique concrète du développement. Les Editions Ouvrières, Paris

Leopold A (1949) A Sand County almanac. Oxford University Press, Oxford

Light A, Katz, E (eds) (1996) Environmental Pragmatism. Routledge, London & New York

Lockwood JA (2004) Locust: the devastating rise and mysterious disappearance of the insect that shaped the American frontier. Basic Books, New York

Long PO (2000) Technology, society, and culture in late medieval and Renaissance Europe, 1300–1600. Society for the History of Technology and the American Historical Association, Washington, DC, pp 7–14

Long PO (2003) Technology and society in the medieval centuries: Byzantium, Islam, and the West, 500–1300. Society for the History of Technology and the American Historical Association, Washington, DC, pp 9–45

Lund V (2006) Natural living – a precondition for animal welfare in organic farming. Livestock Science 100: 71–83

Lyson T (2002) Advanced agricultural biotechnologies and sustainable agriculture. Trends in Biotechnology 20 (5): 193–198

Macer DRJ, Bhardwaj M, Maekawa F, Niimura Y (2003) Ethical opportunities in global agriculture, fisheries, and forestry: the role for FAO. Journal of Agricultural and Environmental Ethics 16: 479–504

MacNeish, RS (1992) The origins of agriculture and settled life. University of Oklahoma Press, Norman, OK

Marcus E (1998) Vegan: the new ethics of eating. McBooks Press, Ithaca, NY

Martinez A (2003) Harvesting poverty. New York Times. http://www.nytimes.com/ref/opinion/harvesting-poverty.html?pagewanted = all

Marx K (1906) Capital, a critique of political economy: the process of capitalist production. Translated from the third German edition by S. Moore and E. Aveling; edited by F. Engels; revised and amplified according to the fourth German edition by E. Untermann. Modern Library, New York

Masood E (1998) Social equity versus private property: striking the right balance. Nature 392: 537

Mayer A, Mayer J (1974) Agriculture, the island empire. Daedelus 103: 83–95

Mayer JE (2005) The Golden Rice controversy: useless science or unfounded criticism. BioScience 55 (9) (September): 726–727

McGlone JJ (1993) What is animal welfare? Journal of Agricultural and Environmental Ethics 6 (Supplement 2): 26–36

Melander AL (1914) Can insects become resistant to sprays? Journal of Economic Entomology 7: 167–172
Merchant C (1980) The death of nature. Harper & Row, San Francisco, CA
Miele M, Murdoch J (2002) Fast food/slow food: standardising and differentiating cultures of food. In: Almas R (ed) Globalization, localization and sustainable livelihoods. Ashgate, Aldershot, UK
Mill JS (2000) On Liberty. In: Three Essays, Oxford University Press, Oxford, p 92
Mkangi K (1993) The African debt crisis: a radical human perspective. ACLCA Publishers, Nairobi, p 10
Moberg GP (1985) Biological response to stress: key to assessment of animal well-being? In: Moberg GP (ed) Animal stress. American Physiological Society, Bethesda, MD, pp 27–49
Nicolas F, Valceschini E (1995) Agro-alimentaire: une economie de la qualité. INRA and Economica, Paris
Norton B (1991) Towards unity among environmentalists. Oxford University Press, Oxford
Nuffield Council of Bioethics (1999) Genetically modified crops: the ethical and social issues. Latimer Trend & Company, London, p 10
Nussbaum MC (2003) Capabilities as fundamental entitlements: Sen and social justice. Feminist Economist 9 (2–3): 34
Oldfield A (2000) Liberal democratic theory: some reflections on its history and its present. In Garrard J, Tolz V, White R (eds) European democratization since 1800. St. Martin's Press, New York, pp 6–25
Olsson IAS, Gamborg C, Sandøe P (2006) Taking ethics into account in farm animal breeding: what can the breeding companies achieve? Journal of Agricultural and Environmental Ethics 19: 37–46
O'Neill J (1993) Ecology, policy, politics. Routledge, London
Oruka HO (1997) Practical Philosophy in Search of Ethical Minimum. East African Education Publishers, Nairobi, p 84–88
O'Toole JC, Toenniessen GH, Harris RR, Herdt RW (2001) The Rockefeller Foundation's international program in rice biotechnology. In: Khush GS, Brar, Hardy B (eds) Rice Genetics IV, Proceedings of the Fourth International Rice Genetics Symposium, 22–27 October 2000. International Rice Research Institute, Los Baños, pp 39–60
Pacey A (2001) Meaning in Technology. MIT Press, Cambridge, MA
Parfit D (1984) Reasons and persons. Oxford University Press, Oxford
Parfit D (1991) Future generations, further problems. Philosophy and Public Affairs 11: 113–172
Peake H (1928) The origins of agriculture. Ernest Benn Limited, London. Reprint (1979). AMS Press, New York
Penman D (1996) The price of meat. Victor Gollancz, London
Perkins JH (1982) Insects, experts, and the insecticide crisis: the quest for new pest management strategies. Plenum, New York
Perkins JH (1997) Geopolitics and the green revolution: wheat, genes, and the Cold War. Oxford University Press, New York
Perkins JH (2003) Engineering the earth and changing scientists: the Rockefeller Foundation and the strategy of agricultural research. Delivered to the Society for the History of Technology, Atlanta, GA (October)
Pogge T (2002) World poverty and human rights. Blackwell, Malden, MA, p 131
Polanyi K (1944) The great transformation. Farrar & Rinehart, New York
Pollard S (1968) The idea of progress: history and society. Penguin Books, Harmondsworth, UK
Potrykus I (2001) The 'golden rice' tale. Vitro Cellular and Developmental Biology-Plants 37: 93–100
Preece R, Fraser D (2000) The status of animals in Biblical and Christian thought: a study in colliding values. Society and Animals 8: 245–263
Rapley J (2002) Understanding development, theory and practice in the third world. Lynne Rienner, Boulder, CO/London
Rawls J (1993) Political liberalism. Columbia University Press, New York, pp 131–172

Reardon T, Berdegue JA (2002) The rapid rise of supermarkets in Latin America: challenges and opportunities for development. Development Policy Review 20: 371–388

Reardon T, Codron J-M, Busch L, Bingen J, Harris CK (2001) Global change in agrifood grades and standards: agribusiness strategic responses in developing countries. International Food and Agribusiness Management Review 2: 421–435

Redclift M (1993) Sustainable development: needs, values, rights. Environmental Values 2 (1): 3–20

Regan T (2004) The case for animal rights. 2nd edn. University of California Press, Berkeley, CA

Ripper WE (1956) Effect of pesticides on balance of arthropod populations. Annual Review of Entomology 1: 403–438

Robbins J (1987) Diet for a new America. Stillpoint Publishing, Walpole, MA

Robertson GP, Broome JC, Chornetsky EA, Frankenberger JR, Johnson P, Lipson M, Miranowski JA, Owens ED, Pimentel D, Thrupp LA (2004) Rethinking the vision for environmental research in US agriculture. BioScience 54: 61–65

Robinett J (1994) This rough magic: technology in Latin American fiction. Peter Lang, New York, pp 3–5, 268

Rogers B (2003) Beef and liberty, roast beef, John Bull and the English nation. Chatto & Windus, London

Roling N (2000) Gateway to the global garden: beta/gamma science for dealing with ecological rationality. Eighth Annual Hopper Lecture, University of Guelph, Canada

Rollin B (1997) On telos and genetic engineering. In: Holland A, Johnson A (eds) Animal biotechnology and ethics. Kluwer, Amsterdam, pp 156–171

Rollin BE (1993) Animal welfare, science, and value. Journal of Agricultural and Environmental Ethics 6 (Supplement 2): 44–50

Rolston H (1985) Duties to endangered species. Bioscience 35: 718–726

Rolston H (1994) Conserving natural value. Columbia University Press, New York

Rolston H (2003) Environmental ethics. In: Bunnin N, Tsui-James EP (eds) The Blackwell companion to philosophy. 2nd edn. Blackwell, Oxford, pp 517–530

Rosenau JN (2003) Distant proximities: dynamics beyond globalization. Princeton University Press, Princeton, NJ

Rosset P (2000) Lessons from green revolution. Institute for Food and Development Policy, p 7. http://foodfirst.org/opeds/2000/4-greenrev.htlm

Rostow WW (1960) The stages of economic growth . A non-communist manifesto. University Press, Cambridge, MA

Rostow WW (1987) Rich countries and poor countries: reflections on the past, lessons for the future. Westview Press, Boulder, CO

Roussel JS, Clower D (1955) Resistance to the chlorinated hydrocarbon insecticides in the boll weevil (Anthonomus grandis Boh.). Louisiana Experiment Station Circular No. 41, Louisiana State University and Agricultural and Mechanical College, September 1955

Rovane C (1998) The bounds of agency: an essay in revisionary metaphysics. Princeton University, Princeton, NJ

Rozin P, Fischler C, Imada S, Sarubin A, Wrzesniewski A (1999) Attitudes to food and the role of food in life in the U.S.A., Japan, Flemish Belgium and France: possible implications for the diet–health debate. Appetite 33: 163–180

Russell E (2001) War and nature, fighting humans and insects with chemicals from World War I to Silent Spring. Cambridge University Press, Cambridge, pp 21, 83

Russell EJ (1966) A history of agricultural science in Great Britain, 1620–1954. George Allen & Unwin, London, pp 110–112

Rutgers B, Heeger R (1999) Inherent worth and respect for animal integrity. In: Dol M, van Vlissingen MF, Kasanmoentalib S, Visser T, Zwart H (eds), Recognizing the intrinsic value of nature. Van Gorcum, Assen, NL, pp 41–53

Sachs J (2005) The end of poverty: economic possibilities for our time. Penguin Press, New York

Sachs W (ed) (1992) The development dictionary, a guide to knowledge as power. Zed Books, London/Atlantic Highlands, NJ

Sainsbury D (1986) Farm animal welfare: cattle, pigs and poultry. Collins, London
Salvadori M (1977) The liberal heresy, origins and historical development. St. Martin's Press, New York, pp 54–56
Sandøe P, Simonsen HB (1992) Assessing animal welfare: where does science end and philosophy begin? Animal Welfare 1: 257–267
Sandøe P, Nielsen BL, Christensen LG, Sørensen P (1999) Staying good while playing God - the ethics of breeding farm animals. Animal Welfare 8 (4): 313–328
Savory CJ (2004) Laying hen welfare standards: a classic case of 'power to the people.' Animal Welfare 13 (Supplement 1): 153–158
Schech S, Haggis J (2000) Culture and development, a critical introduction. Blackwell, Oxford
Scherer D (2003) The ethics of sustainable resources. In: Light A, Rolston H (eds) Environmental ethics. Blackwell, Oxford
Schroder M, McEachem M (2004) Consumer value conflicts surrounding ethical food purchase decisions: a focus on animal welfare. International Journal of Consumer Studies 28 (2): 168–178
Schultz TW (1972) Knowledge, agriculture and welfare. Science Studies 2 (4): 367
Schweitzer A (1923) The philosophy of civilization. Reprint (1987). Prometheus, Buffalo, NY
Seidel G (1998) Biotechnology in animal agriculture. In: Holland A, Johnson A (eds) Animal biotechnology and ethics. Kluwer, Amsterdam, pp 51–68
Sen A (1999) Development as freedom. Anchor Books, New York
Sequoia A (1990) 67 ways to save the animals. Harper Collins, New York
Shiva V (2005) How to end poverty: making poverty history and the history of poverty. ZNet Daily Commentaries. http://www.zmag.org/sustainers/content/2005–05/11shiva.cfm
Shue H (1980) Basic Rights: Subsistence, Affluence, and U.S. Forcign Policy. Princeton University Press, Princeton. p19
Singer P (1990) Animal liberation. Revised edn. Avon Books, New York
Singer P (1993) Practical ethics. 2nd edn. Cambridge University Press, Cambridge
Smith A (1776) The wealth of nations. Republished (1904). Dent, London
Sommer A (2001) Vitamin A deficiency disorders: origins of the problem and approaches to its control. http://www.rockfound.org. Accessed 6 September 2003
Statistics Canada (2005) Census of agriculture 2004 (data updated 21 January 2005). http://www.statcan.ca/english/Pgdb/agrc33a.htm
Stenmark M (2002) Environmental ethics and policy making. Ashgate, Aldershot, UK
Stevenson P (2004) European Union law on the welfare of farm animals. Compassion in World Farming Trust, Petersfield, UK. http://www.ciwf.org.uk/publications/reports/EU_Law_2004.pdf. Accessed November 2004
Stiglitz JE (2002) Globalization and its discontents. W.W. Norton, New York/London
Strange M (1988) Family farming: a new economic vision. Institute for Food and Development Policy, San Francisco, CA
Streeten P (1981) First things first. Oxford University Press, New York
Streeten P (1994) Strategies for human development: global poverty and unemployment. Handelshojskolens Forlag, Copenhagen
Sustainable Table and Free Range Studios (2003) The Meatrix. http://www.themeatrix1.com/. Accessed October 2007
Tannenbaum J (1991) Ethics and animal welfare: the inextricable connection. Journal of the American Veterinary Medical Association 198: 1360–1376
Tanner B (2000) Independent assessment by third-party certification bodies. Food Control 11: 415–417
Taylor GB (1972) One man's philosophy of welfare. Veterinary Record 91: 426–428
Taylor PW (1986) Respect for nature: a theory of environmental ethics. Princeton University Press, Princeton, NJ
Tenner E (1996) Why things bite back: technology and the revenge of unintended consequences. Vintage, New York

te Velde H, Aarts N, van Woerkum C (2002) Dealing with ambivalence: farmers' and consumers' perceptions of animal welfare in livestock breeding. Journal of Agricultural and Environmental Ethics 15: 203–219

Thiermann AB, Babcock S (2005) Animal welfare and international trade. Revue Scientifique et Technique de l'Office International des Epizooties 24: 747–755

Thompson PB (1995) The spirit of the soil: agriculture and environmental ethics. Routledge, London

Thompson PB (1998) Agricultural ethics – research, teaching, and public policy. Iowa State University Press, Ames, IA

Thompson PB (2001) The reshaping of conventional farming: a North American perspective. Journal of Agricultural and Environmental Ethics 14: 217–229

Thompson PB (2004) Getting pragmatic about farm animal welfare. In: McKenna E, Light A (eds) Animal pragmatism: rethinking human-nonhuman relationships. Indiana University Press, Bloomington, IN, pp 140–159

Thompson PB (2007) Food biotechnology in ethical perspective. 2nd edn. Springer, Dordrecht, The Netherlands

Thoreau HD (1854 [1995 edn]) Walden, or, life in the woods. Dover, New York

Thornton G (2003) WATT Poultry USA's rankings: broilers – turbulent 2002 leads to rationalization. http://www.wattnet.com/ (subscription required). Accessed February 2005

Toenniessen GH (2000) Vitamin A deficiency and golden rice: the role of the Rockefeller Foundation. http://www.rockfound.org. Accessed 6 September 2003

Truth and Reconciliation Commission of South Africa Report (1998), Vol. 4. Truth and Reconciliation Commission, Cape Town

Tudge C (1998) Neanderthals, bandits and farmers, how agriculture really began. Yale University Press, New Haven, CT

United Nations (1948) Article 27 in All Human Rights for All. Fiftieth Anniversary of the Universal Declaration of Human Rights 1948–1998. Adopted and proclaimed by General Assembly resolution 217 A (III) of 10 December 1948. http://www.un.org/Overview/rights.html. Accessed 29 January 2008

United Nations Development Program (2003) Human development report 2003. Millenium development goals: a compact among nations to end human poverty. Oxford University Press, New York/Oxford

United States Department of Agriculture, Economic Research Service (1969) DDT used in farm production. Agricultural economic report no. 188, Washington, DC

United States Department of Agriculture, Economic Research Service (2003) United States Fact Sheet

United States Department of Agriculture, Economic Research Service (2004) Poultry Yearbook. USDA, Washington, DC. http://www.ers.usda.gov/data/sdp/view.asp?f = livestock/89007/)

Vaarst M, Roderick S, Lund V, Lockertz W, Hovi M (2004) Organic principles and values: the framework for organic animal husbandry. In: Vaarst M, Roderick S, Lund V, Lockertz W (eds) Animal health and welfare in organic agriculture. CAB International, Wallingford, UK, pp 1–12

Vandermeer J (2003) Agricultural scientists are wrong about agriculture and conservation scientists are wrong about conservation. Endangered Species Update 20: 53

van der Ploeg JD (1990) Labor, markets, and agricultural production. Westview Press, Boulder, CO

Varner G (1998) In nature's interests? Oxford University Press, New York

Vorstenbosch J (1993) The concept of integrity. Its significance for the ethical discussion on biotechnology and animals. Livestock Production Science 36: 109–112

Waiblinger S, Menke C (1999) Influence of herd size on human–cow relationships. Anthrozoos 12: 240–247

Waiblinger S, Baumgartner J, Kiley-Worthington M, Niebuhr K (2004) Applied ethology: the basis for improved animal welfare in organic farming. In: Vaarst M, Roderick S, Lund V, Lockeretz W (eds) Animal health and welfare in organic agriculture. CAB International Publishing, Wallingford, UK, pp 117–161

Warren MA (2000) On the moral and legal status of abortion. In: Lafollette H (ed) Ethics in practice. Blackwell, Oxford/Washington, DC, pp 87–99

Webster J (1994) Animal welfare: a cool eye towards Eden. Blackwell, Oxford

Welch RM, Graham RD (1990) A new paradigm for world agriculture: meeting human needs, productive and nutritious. Field Crops Research 60 (1990): 1–10

Wenz P (1993) Minimal, moderate, and extreme moral pluralism. Environmental Ethics 15: 61–74

Winter M (2004) Geographies of food: agro food geographies – farming, food and politics. Progress in Human Geography 28 (5): 664–670

World Commission on Environment and Development (1987) Our common future. Oxford University Press, Oxford

World Development Report (2002) Building institutions for markets. Oxford University Press, New York

Worster D (1994) Nature's economy: a history of ecological ideas. 2nd edn. Cambridge University Press, New York, pp 2–55

Ye X, Al-Babili S, Klöti A, Zhang J, Lucca P, Beyer P, Potrykus I (2000) Engineering the provitamin A (β-carotene) biosynthetic pathway into (carotenoid-free) rice endosperm. Science 287 (14 January): 303–305

Zimdahl R (2006) Agriculture's ethical horizon. Academic, New York

Index

Printed in the United Kingdom
by Lightning Source UK Ltd.
133928UK00002B/94-111/P

9 781402 087219